CliffsNotes® Statistics QuickReview

CliffsNotes® Statistics QuickReview

By David H. Voelker, M.A., Peter Z Orton, Ed.M., Scott V. Adams, Ph.D., and John J. Diehl, M.S.

About the Authors

David H. Voelker is an educator from the San Francisco Bay area. Peter Z Orton earned his Ph.D. from Stanford University. Scott V. Adams earned his Ph.D. from Vanderbilt University. John J. Diehl earned his M.S. in mathematics from Drake University.

Publisher's Acknowledgments

Editorial

Acquisitions Editor: Greg Tubach
Project Editor: Elizabeth Kuball
Copy Editor: Elizabeth Kuball
Technical Editors: Mike McAsey, Mary Jane Sterling

Composition

Indexer: BIM Indexing & Proofreading
Proofreader: Henry Lazarek
Wiley Publishing, Inc. Composition Services

www.learneo.com

Library of Congress Cataloging-in-Publication data is available.
ISBN 978-1-957671-17-8

Printed in the United States of America
10 9 8 7 6 5 4 3 2 1

Table of Contents

INTRODUCTION

Statistics are a powerful tool for finding patterns in data and inferring important connections between events in the real world. We are accustomed to seeing statistics every day—in the news, for example. However, many people find statistics confusing and intimidating. Often, statistics are considered to make an argument irrefutable. On the other hand, some people perceive statistics as a trick, used to support outrageous claims.

This book is intended to clarify some of the elements of statistical reasoning and analysis. The format is simple, and there are many examples to help translate abstract ideas into concrete applications.

Why You Need This Book

Can you answer yes to any of these questions?

- Do you need to review the fundamentals of statistics quickly?
- Do you need a course supplement to statistics?
- Do you need a concise, comprehensive reference for statistics?
- Do you need to prepare for a statistics exam?

If so, then *CliffsNotes Statistics Quick Review,* 2nd Edition, is for you!

How to Use This Book

You are in charge here. You get to decide how to use this book. You can read it cover to cover or just look up the information you need right now. Here are some recommended ways to search for topics:

- Flip through the book looking for subject areas at the top of each page.
- Get a glimpse of what you'll gain from a chapter by reading through the "Chapter Check-In" at the beginning of each chapter.

- Use the "Chapter Check-Out" at the end of each chapter to gauge your grasp of the important information you need to know.
- Test your knowledge more completely in the Review and look for additional sources of information in the Resource Center.
- Use the Glossary to find key terms quickly. This book defines new terms and concepts where they first appear in the chapter. If a word is boldfaced, you can find a more complete definition in the book's glossary.
- Flip through the book until you find what you're looking for—we organized this book to gradually build on key concepts.

Hundreds of Practice Questions Online!

Prepare for your next quiz or test with hundreds of additional practice questions online. The questions are organized by this book's chapter sections, so it's easy to use the book and then quiz yourself online to make sure you know the subject. Go to www.cliffsnotes.com/go/quiz/statistics to test yourself anytime and find other free homework help.

Chapter 1
OVERVIEW

Chapter Check-In

- ❑ Learning about the different kinds of statistics
- ❑ Learning about variables
- ❑ Understanding the steps in statistical reasoning
- ❑ Learning several uses for statistical reasoning

For many people, statistics means numbers—numerical facts, figures, or information. Statistics is about **data.** Data consists of information about statistical **variables.** There are two types of variables: **Quantitative variables** are variables that can be measured or described by values, such as height. Categorical variables have values that are categories, such as type of pet. The data for these variables are usually counts or frequencies of the numbers for each category.

Reports of industry production, baseball batting averages, government deficits, and so forth, are often called statistics. To be precise, these numbers are **descriptive statistics** because they are numerical data that describe phenomena. Descriptive statistics are as simple as the number of children in each family along a city block or as complex as the annual report released from the U.S. Department of the Treasury.

Types of Statistics

In these pages, we are concerned with two ways of representing descriptive statistics: numerical and pictorial.

Numerical statistics

Numerical statistics are numbers, but clearly, some numbers are more meaningful than others. For example, if you are offered a purchase price of $1 for an automobile on the condition that you also buy a second automobile, the price of the second automobile would be a major consideration (its price could be $1,000,000 or $1,000); thus, the average—or **mean**—of the two prices would be the important statistic. (The different meaningful numerical statistics are discussed in Chapter 3.)

Pictorial statistics

Taking numerical data and presenting it in pictures or graphs is known as **pictorial statistics.** Showing data in the form of a graphic can make complex and confusing information appear more simple and straightforward. Different types of graphs are used for quantitative and categorical variables. (The more commonly used graphics in statistics are discussed in Chapter 2.)

Method of Statistical Inference

Statistics is also a method, a way of working with numbers to answer puzzling questions about both human and nonhuman phenomena. Questions answerable by using the "method" of statistics are many and varied: Which of several techniques is best for teaching reading to third-graders? Will a new medicine be more effective than the old one? Can you expect it to rain tomorrow? What is the probable outcome of the next presidential election? Which assembly-line process produces fewer faulty carburetors? How can a polling organization make an accurate prediction of a national election by questioning only a few thousand voters? And so on.

For our purposes, statistics is both a collection of numbers and/or pictures *and* a process: the art and science of making accurate guesses about outcomes involving numbers.

So, fundamentally, the goals of statistics are

- To describe variables and data
- To make accurate inferences about groups based upon incomplete information

Steps in the Process

Making accurate guesses requires groundwork. The statistician must do the following in order to make an educated, better-than-chance hunch:

1. Gather data (numerical information).
2. Organize the data (sometimes in a pictorial).
3. Analyze the data (using tests of significance and so forth).

This book shows you how to do precisely these procedures and how to use your analysis to draw an **inference**—an educated statistical guess—to solve a particular problem. While these steps may appear simple (and indeed some of them are), sound statistical method requires that you perform them in certain prescribed ways. Those ways are the heart and soul of statistics.

Making Predictions

Suppose that you decide to sell commemorative T-shirts at your town's centennial picnic. You know that you can make a tidy profit, but only if you can sell most of your supply of shirts because your supplier will not buy any of them back. How many shirts can you reasonably plan on selling?

Your first question, of course, is this: How many people will be attending the picnic? Suppose you know that 100,000 tickets to the event have been sold. How many T-shirts should you purchase from your supplier to sell on the day of the event? 10,000? 50,000? 100,000? 150,000? How many of each size—small, medium, large, extra-large? And the important question: How many T-shirts must you sell in order to make some profit for your time and effort and *not* be left with an inventory of thousands of unsold shirts?

Ideally, *before* you buy your inventory of T-shirts, you need to have an accurate idea of just how many ticket holders will want to purchase centennial T-shirts and which sizes they will want. But, obviously, you have neither the time nor the resources to ask all 100,000 people whether they plan to purchase commemorative T-shirts. If, however, you could locate a small number of those ticket holders—for example, 100—and get an accurate count of how many of those 100 would purchase a T-shirt, you would have a better idea of how many of the 100,000 ticket holders would be willing to buy one.

That is, of course, if the 100 ticket holders that you asked (called the **sample**) are not too different in their intentions to purchase T-shirts from the total 100,000 ticket holders (called the **population**). If the sample is indeed representative (typical) of the population, you could expect about the same percentage of T-shirt sales (and sizes) for the population as for the sample, all things being equal. So, if 50 of your sample of 100 people say they cannot wait to plunk down $10 for a centennial T-shirt, it would be reasonable to expect that you would sell about 50,000 T-shirts to your population of 100,000. (At just $1 profit per T-shirt, that is $50,000!)

But before you start shopping to buy a yacht with your profits, remember that this prediction of total T-shirt sales relies heavily upon the sample being representative (similar to the population), which may not necessarily be the case with your sample. You may have inadvertently selected a sample that has more expendable income or is a greater proportion of souvenir T-shirt enthusiasts or who knows what else. Are you reasonably certain that the intentions of the sample of 100 ticket holders reflect the intentions of the 100,000 ticket holders? If not, you may quite possibly be stuck with tens of thousands of centennial T-shirts and no profit to splurge on a yacht.

You can see why choosing a random sample is a critical part of the process of statistics and why this book includes a chapter on sampling (Chapter 5). Even with careful sampling methods, our conclusions are still educated guesses. This is because one sample does not perfectly represent the population, and different samples may give different results.

Comparing Results

Making predictions is only one use of statistics. Suppose you have recently developed a new headache/pain remedy that you call Ache-Away. Should you produce Ache-Away in quantity and make it available to the public? That would depend upon, among other factors, whether Ache-Away is more effective than the old remedy. How can you determine that?

One way might be to administer both remedies to two separate groups of people, collect data on the results, and then statistically analyze that data to determine if Ache-Away is more effective than the old remedy. And what if the results of this test showed Ache-Away to be more effective? How certain can you be that *this* particular test administration is indicative of *all* tests of these two remedies? Perhaps the group taking the old remedy (the control group) and the group taking Ache-Away (the treatment group) were so dissimilar that the results were due *not* to the pain remedies but to the differences between the groups.

It is possible that the results of this test are far off the results that you would get if you tried the test several more times. You certainly do not want to foist a questionable drug upon an unsuspecting public based upon atypical test results. How certain can you be that you can put your faith in the results of your tests? You can see that the problems of comparing headache remedy results can produce headaches of their own.

This type of statistical procedure is discussed in Chapters 6 and 7.

Probability

One of the most familiar uses of statistics is to determine the chance of some occurrence. For instance, what are the chances that it will rain tomorrow or that the Chicago Cubs will win a World Series? These kinds of probabilities, although interesting, are not the variety under discussion here. Rather, we are examining the probability in statistics that deals with classic theory and frequency theory—events that can be repeated over and over again, independently, and under the same conditions.

Coin tossing and card drawing are two such examples. A *fair coin* (one that is not weighted or *fixed*) has an equal chance of landing heads as landing tails. A typical deck of cards has 52 different cards—13 of each suit (hearts, clubs, diamonds, and spades)—and each card or suit has an equal chance of being drawn. This kind of event forms the basis of your understanding of probability and enables you to find solutions to everyday problems that seem far removed from coin tossing or card drawing.

Probability helps us describe our conclusions in a way that takes the uncertainty from random sampling into account. (Probability is discussed in Chapter 4.)

Common Statistics Mistakes

Whether you are new to the study of statistics or you have been using statistical analysis for years, you likely will make (or already have made) errors in your application of principles explained in this book. In fact, some particular errors are committed so frequently in statistical analysis that we feel obliged to call them to your attention in the hopes that, by being prepared for them, you will avoid their pitfalls. Appendix A lists these common mistakes.

Chapter Check-Out

1. True or False: Statistics are about things only in the past, so you cannot use them to make predictions.
2. True or False: An inference is an educated guess, made after analyzing data.
3. True or False: A population is a randomly chosen portion of a larger sample.
4. True or False: Statistical analysis will allow you to be absolutely sure of your conclusions.

Answers: 1. False **2.** True **3.** False **4.** False

Chapter 2
GRAPHIC DISPLAYS

Chapter Check-In

- ❑ Displaying categorical data in simple graphical formats such as bar charts and pie charts
- ❑ Displaying quantitative variable data in simple graphical formats such as dot plots, frequency histograms, and stem-and-leaf plots
- ❑ Using box plots to display numerical measures of data
- ❑ Interpreting graphic displays to make conclusions about the distribution of the variable
- ❑ Understanding scatter plots

Pie charts and bar charts are graphic displays of data for categorical variables. Dot plots, stem-and-leaf plots, histograms, and box-and-whisker plots are graphic displays of data for numerical variables. This chapter discusses how to construct these graphic displays.

As an example, consider the yearly expenditures of a college undergraduate. After collecting her **data** (expense records) for the past year, she finds the expenditures shown in Table 2-1.

Table 2-1 Yearly Expenses of College Undergraduate

Item	*Amount*
Tuition fees	\$5,000
Room and board	\$9,000
Books and lab	\$2,000
Clothes/cleaning	\$1,000
Transportation	\$2,000
Insurance and miscellaneous	\$1,000

These figures, although presented in categories, do not allow for easy analysis. The reader must expend extra effort in order to compare amounts spent or relate individual proportions to the total. For ease of analysis, these data can be presented pictorially.

Bar Chart

One way to pictorially display the numbers shown in Table 2-1 is with a **bar chart** (see Figure 2-1).

Figure 2-1 Vertical bar chart presentation of the expenditures of a college undergraduate for the past year.

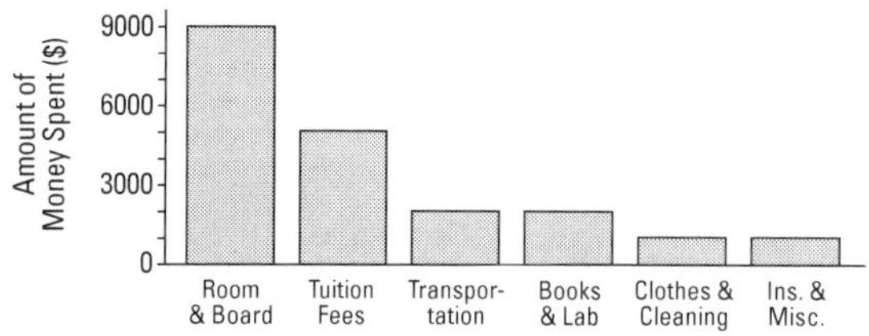

Comparing the size of the bars, you can quickly see that room-and-board expenses are nearly double the tuition fees, and tuition fees are more than double the expenses for books and lab and transportation.

A bar chart also may be placed on its side with the bars going horizontally, as shown in Figure 2-2.

Figure 2-2 Horizontal bar chart presentation of the expenditures of a college undergraduate for the past year.

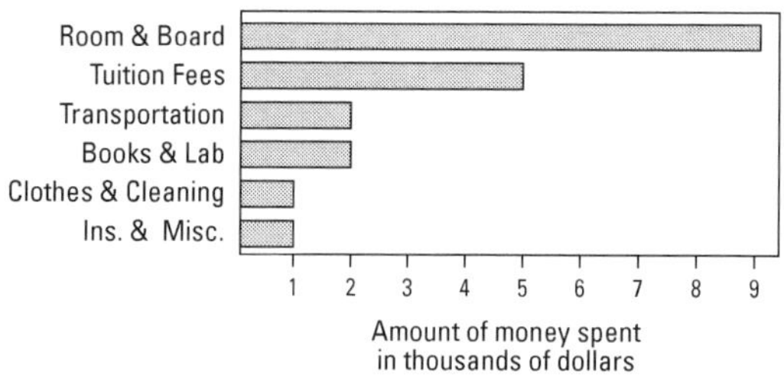

In each bar chart, vertical or horizontal, the amounts are ordered from highest to lowest or lowest to highest, making the chart clearer and easier to understand. Space is left between each of the bars in order to define the categories as being different.

The bottom line in Figure 2-1 and the left side in Figure 2-2 indicate 0. Although typical, this presentation need not always be used. Finally, although the lengths of the bars may be different, their thicknesses are the same.

Pie Chart

Bar charts have a limitation: It is difficult to see what portion of the total each item comprises. If knowing about a part of the whole is an important consideration, a **pie chart** is a better choice for showing the same data. A pie chart also may display each category's percentage of the total. Using the same data from the undergraduate expenditures, we get the pie chart shown in Figure 2-3.

Figure 2-3 Pie chart presentation of the expenditures of a college undergraduate for the past year.

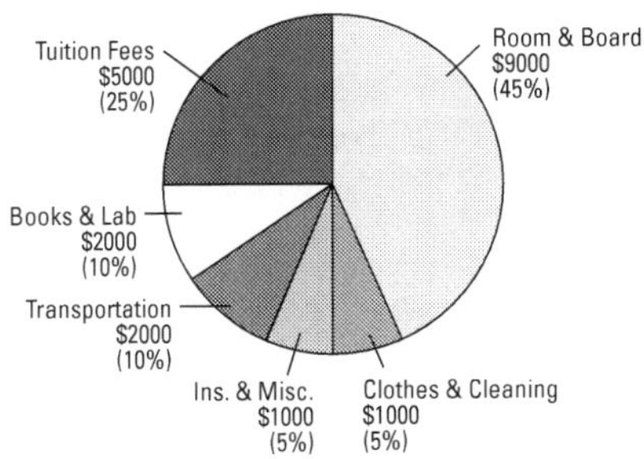

The parts of the circle (or *pie)* match in size each category's percentage of the total. The parts of the pie chart are ordered from highest to lowest for easier interpretation of the data. Pie charts work best with only a few categories; too many categories make a pie chart confusing.

Dot Plot

Dot plots are used for quantitative variables. Typically used for a small set of values, a dot plot uses a *dot* for each unit of measurement. For the preceding undergraduate expense data, if we just wanted to compare the dollar values and were not concerned with the categories, a dot plot would look like that shown in Figure 2-4.

Figure 2-4 Dot plot of the expenditures of a college undergraduate for the past year.

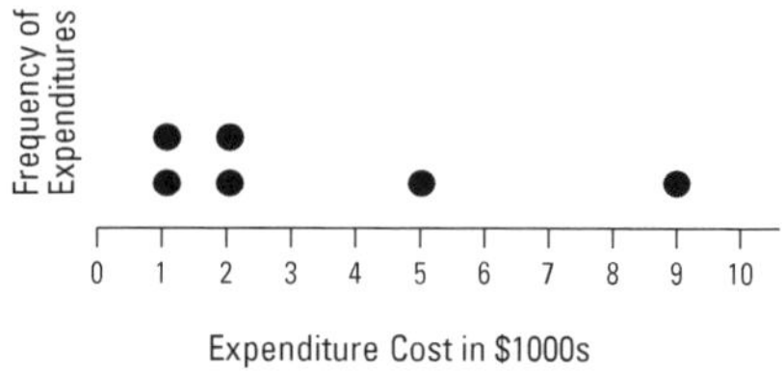

Ogive

Data may be expressed using a single line. An **ogive** (a cumulative line graph) is best used when you want to display the total at any given time. The relative slopes from point to point will indicate greater or lesser increases; for example, a steeper slope means a greater increase than a more gradual slope. An ogive, however, is not the ideal graphic for showing comparisons between categories because it simply combines the values in each category, thus indicating an *accumulation* (a growing or lessening total). If you simply want to keep track of a total and your individual values are periodically combined, an ogive is an appropriate display.

For example, if you saved $300 in both January and April and $100 in each of February, March, May, and June, an ogive would look like Figure 2-5.

Figure 2-5 Ogive of accumulated savings for one year.

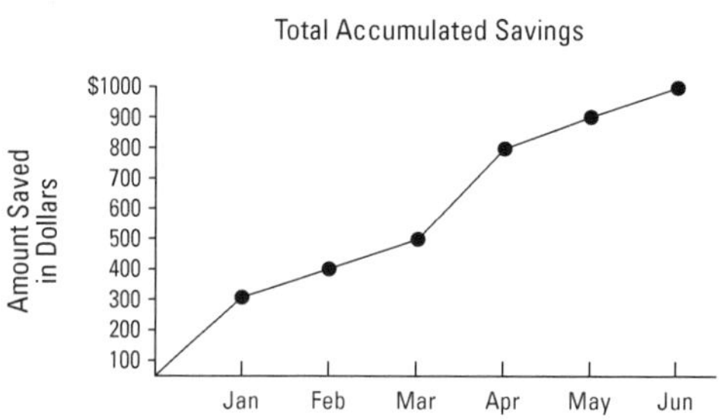

An ogive displays a *running total.* Although each individual month's savings could be expressed in a bar chart (as shown in Figure 2-6), you could not easily see the amount of total growth or loss, as you can in an ogive.

Figure 2-6 Vertical bar chart of accumulated savings for one year.

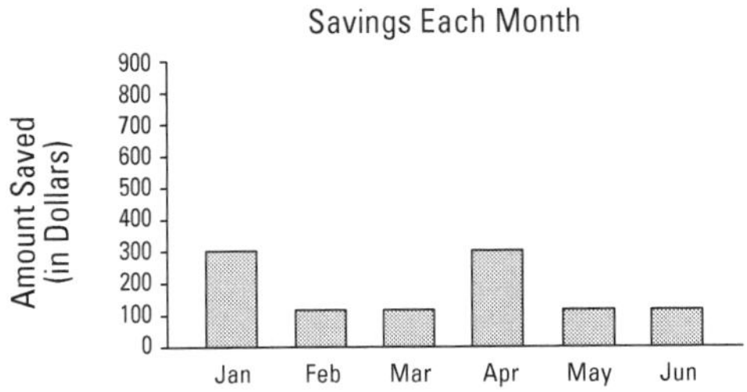

The choice of graphic display, therefore, depends on what information is important for your purposes: percentages (parts of the whole), running total, comparisons of categories, and so forth.

Frequency Histogram

One of the more commonly used pictorials in statistics is the frequency histogram, which in some ways is similar to a bar chart and tells how many items are in each numerical category. For example, suppose that after a garage sale, you want to determine which items were the most popular: the high-priced items, the low-priced items, and so forth. Let's say you sold a total of 32 items for the following prices: \$1, \$2, \$2, \$2, \$5, \$5, \$5, \$5, \$7, \$8, \$10, \$10, \$10, \$10, \$11, \$15, \$15, \$15, \$19, \$20, \$21, \$21, \$25, \$25, \$29, \$29, \$29, \$30, \$30, \$30, \$35, and \$35.

The items sold *ranged* in price from \$1 to \$35. First, divide this **range** of \$1 to \$35 into a number of categories, called **class intervals.** Typically, no fewer than 5 and no more than 20 class intervals work best for a frequency histogram.

Choose the first class interval to include your lowest (smallest value) data and make sure that no *overlap* exists so that one piece of data does not fall into two class intervals. For example, you would not have your first class interval be \$1 to \$5 and your second class interval be \$5 to \$10 because the four items that sold for \$5 would belong in both the first and the second intervals. Instead, use \$1 to \$5 for the first interval and \$6 to \$10 for the second. Class intervals are mutually exclusive.

First, make a table of how your data is distributed (see Table 2-2). The number of observations that falls into each class interval is called the **class frequency.**

Table 2-2 Distribution of Items Sold at Garage Sale

Class	*Interval*	*Frequency*
1	$1 to $5	8
2	$6 to $10	6
3	$11 to $15	4
4	$16 to $20	2
5	$21 to $25	4
6	$26 to $30	6
7	$31 to $35	2

Note that each class interval has the same width. That is, $1 to $5 has a width of five dollars, inclusive; $6 to $10 has a width of five dollars, inclusive; $11 to $15 has a width of five dollars, inclusive; and so forth. From the data, a frequency histogram would look like what you see in Figure 2-7.

Figure 2-7 Frequency histogram of items sold at a garage sale.

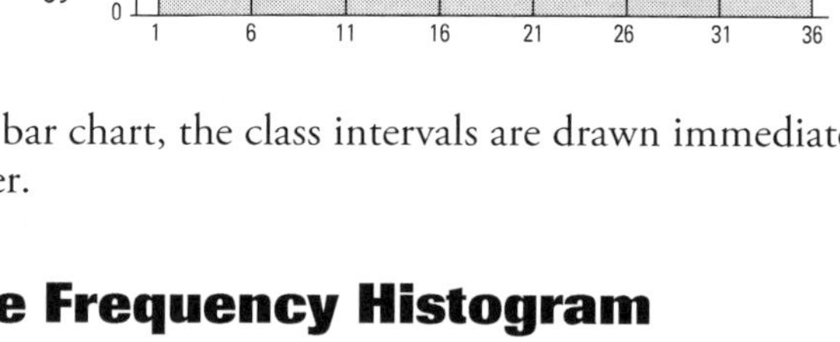

Unlike in a bar chart, the class intervals are drawn immediately adjacent to each other.

Relative Frequency Histogram

A relative frequency histogram uses the same information as a frequency histogram but compares each class interval to the total number of items. For example, the first interval ($1 to $5) contains 8 out of the total of 32 items, so the relative frequency of the first class interval is $\frac{8}{32}$ (see Table 2-3).

Table 2-3 Distribution of Items Sold at Garage Sale, Including Relative Frequencies

Class	*Interval*	*Frequency*	*Relative Frequency*
1	$1 to $5	8	0.25
2	$6 to $10	6	0.1875
3	$11 to $15	4	0.125
4	$16 to $20	2	0.0625
5	$21 to $25	4	0.125
6	$26 to $30	6	0.1875
7	$31 to $35	2	0.0625

The only difference between a frequency histogram and a relative frequency histogram is that the vertical axis uses relative or proportional frequency instead of simple frequency (see Figure 2-8).

Figure 2-8 Relative frequency histogram of items sold at a garage sale.

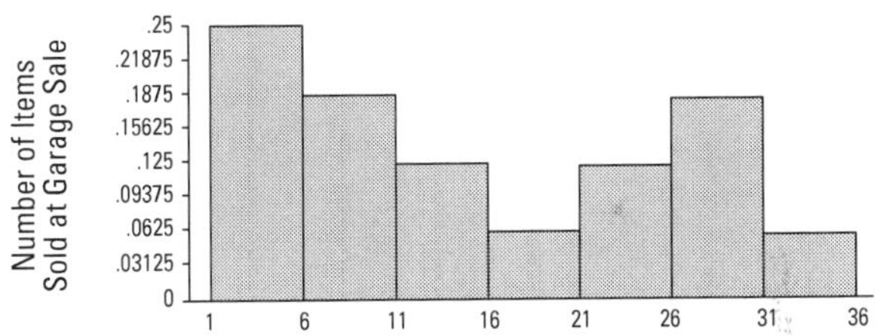

Frequency Polygon

Relative frequencies of class intervals also can be shown in a **frequency polygon.** In this chart, the frequency of each class is indicated by points or dots drawn at the midpoints of each class interval. Those points are then connected by straight lines.

Comparing the frequency polygon (shown in Figure 2-9) to the frequency histogram (refer to Figure 2-7), you see that the major difference is that points replace the bars.

Figure 2-9 Frequency polygon display of items sold at a garage sale.

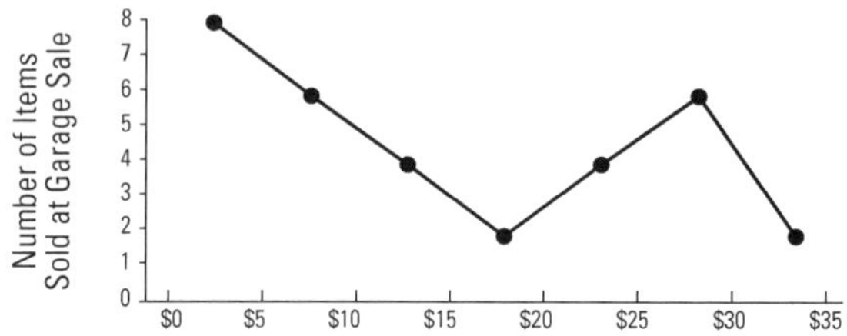

Whether to use bar charts or histograms depends on the data. For example, you may have categorical (or **qualitative)** data—numerical information about categories that vary significantly in kind. Gender (male or female), types of automobile owned (sedan, sports car, pickup truck, van, and so forth), and religious affiliations (Christian, Jewish, Muslim, and so forth) are all qualitative data. On the other hand, **quantitative** data can be measured in amounts: age in years, annual salaries, inches of rainfall. Typically, qualitative data are better displayed in bar charts; quantitative data, in histograms.

Frequency Distribution

Frequency distributions are like frequency polygons (refer to Figure 2-9); however, instead of straight lines, a frequency distribution uses a smooth curve to connect the points and, similar to a graph, is plotted on two axes: The horizontal axis from left to right (or *x*-axis) indicates the different possible **values** of some **variable** (a phenomenon where observations vary from trial to trial). The vertical axis from bottom to top (or *y*-axis) measures frequency or how many times a particular value occurs.

For example, in Figure 2-10, the *x*-axis might indicate annual income (the values would be in thousands of dollars); the *y*-axis might indicate frequency (millions of people or percentage of working population). Notice that in Figure 2-10, the highest percentage of the working population would, thus, have an annual income in the middle of the dollar values. The lowest percentages would be at the extremes of the values: nearly 0 and extremely high.

Figure 2-10 A symmetric bell curve.

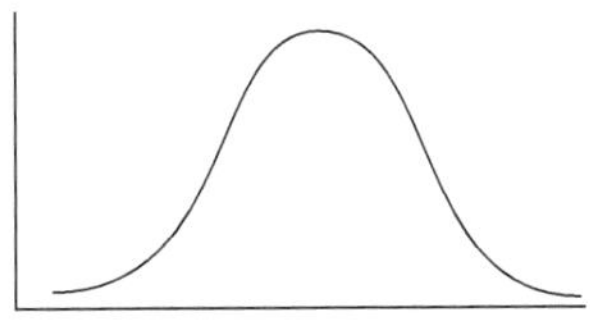

Notice that this frequency curve displays perfect **symmetry;** that is, one half (the left side) is the mirror image of the other half (the right side). Not all frequency curves are perfectly symmetrical.

Figure 2-11 shows a J-shaped curve.

Figure 2-11 J-shaped curve.

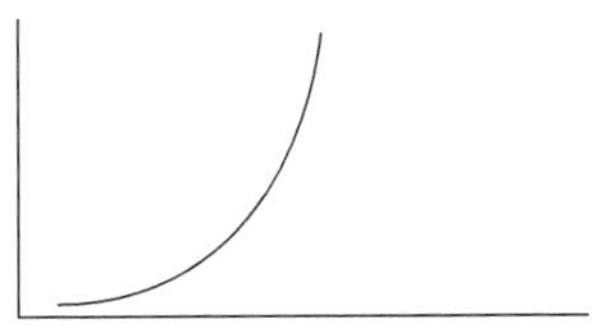

Unlike Figure 2-10, a **bimodal** curve (shown in Figure 2-12) has two high points.

Figure 2-12 A bimodal curve has two maximum peaks.

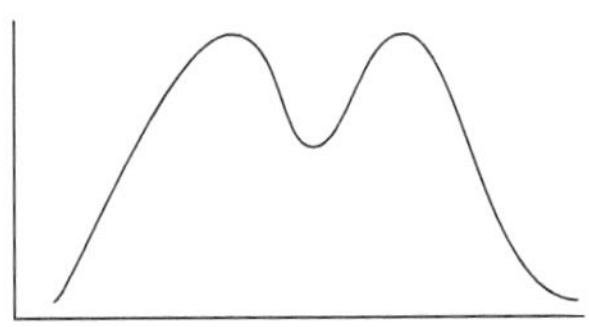

Stem-and-Leaf

Another useful pictorial is the **stem-and-leaf.** It is similar to a histogram in that it shows the range of data, where the data are concentrated, if there are any **outliers** (occasional extremely high or extremely low scores), and the general shape of the distribution.

For example, look at the following data—test scores of 17 high school students: 69, 75, 77, 79, 82, 84, 87, 88, 89, 89, 89, 90, 91, 93, 96, 100, and 100. The stem-and-leaf takes all but the last digit of each score as the stem and uses the remaining digit as the leaf.

As an example, for the score of 69, the 6 is the stem and the 9 is the leaf; for the next three grades (75, 77, and 79), 7 is the stem, and 5, 7, and 9 are the leaves.

Note, too, that along the extreme left side of the chart is a vertical column that keeps a running count or total. (Some stem-and-leafs do not include this running total.) Having a running total enables the reader to quickly locate the median. (Median is discussed further in Chapter 3.)

The completed stem-and-leaf for the high school students' test scores looks like Table 2-4.

Table 2-4 Stem-and-Leaf Display of Students' Test Scores

Running Count	*Stem*	*Leaves*
1	6	9
4	7	5, 7, 9
11	8	2, 4, 7, 8, 9, 9, 9
15	9	0, 1, 3, 6
17	10	0, 0

Notice that, like a histogram, each stem determines a class interval and, also like a histogram, the class intervals are all equal. (In this case, each interval width is from a possible low of 0 to a possible high of 9.) All 17 scores are displayed in the stem-and-leaf so that you can see not only the frequencies and the shape of the distribution but also the actual value of every score.

Box Plot (Box-and-Whiskers)

Box plots, sometimes called **box-and-whiskers,** take the stem-and-leaf one step further. A box plot will display a number of values of a distribution of numbers:

- The median value
- The lower quartile (Q_1)

- The upper quartile (Q_3)
- The interquartile range (*IQR*), the distance between the lower and upper quartiles
- The symmetry of the distribution
- The highest and lowest values

Use the set of values in Table 2-5 to examine each of the preceding items.

Table 2-5 Verbal SAT Scores of 20 High School Students

280	340	440	490	520	540	560	560	580	580
600	610	630	650	660	680	710	730	740	740

The median (the middle value in a set that has been ordered lowest to highest) is the value above which half of the remaining values fall and below which the other half of the remaining values fall. Because there is an even number of scores in our example (20), the median score is the average of the two middle scores (10th and 11th)—580 and 600—or 590.

The **lower quartile** (Q_1 or 25th percentile) is the median of the bottom half. The bottom half of this set consists of the first ten numbers (ordered from low to high): 280, 340, 440, 490, 520, 540, 560, 560, 580, and 580. The median of those ten is the average of the fifth and sixth scores—520 and 540—or 530. The lower-quartile score is 530.

The **upper quartile** (Q_3 or 75th percentile) is the median score of the top half. The top half of this set consists of the last ten numbers: 600, 610, 630, 650, 660, 680, 710, 730, 740, and 740. The median of these ten is again the average of the fifth and sixth scores—in this case, 660 and 680—or 670. So 670 is the upper-quartile score for this set of 20 numbers.

A **box plot** can now be constructed as follows: The left side of the box indicates the lower quartile; the right side of the box indicates the upper quartile; and the line inside the box indicates the median. A horizontal line is then drawn from the lowest value of the distribution through the box to the highest value of the distribution. (This horizontal line is the "whiskers.")

Using the Verbal SAT scores in Table 2-5, a box plot would look like Figure 2-13.

Figure 2-13 A box plot of SAT scores displays median and quartiles.

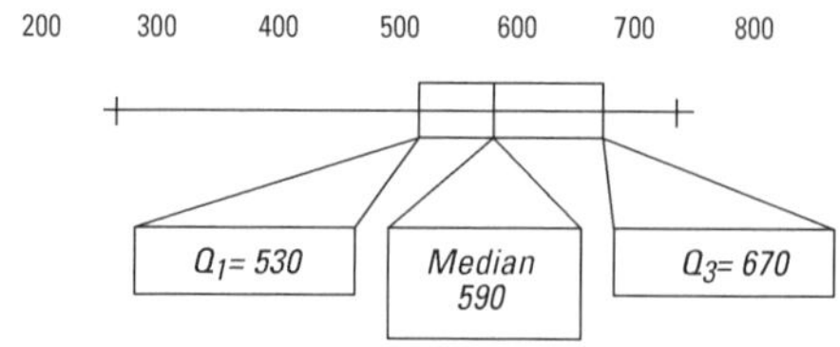

Without reading the actual values, you can see by looking at the box plot in Figure 2-13 that the scores range from a low of 280 to a high of 740; that the lower quartile (Q_1) is at 530; that the median is at 590; and that the upper quartile (Q_3) is at 670. Because the median is slightly nearer the lower quartile than the upper quartile and the interquartile range is situated far to the right of the range of values, the distribution departs from symmetry.

Scatter Plot

Sometimes you want to display information about the relationship involving two different phenomena. These data would be called **bivariate** or paired data. For example, suppose you collected data about the number of days that law-school candidates studied for a state bar examination and their resulting scores on the exam. The data from eight candidates is shown in Table 2-6.

Table 2-6 Law School Candidates' Prep Times and Test Scores

Candidate	***1***	***2***	***3***	***4***	***5***	***6***	***7***	***8***
Days studied	7	9	5	1	8	4	3	6
Score earned	23	25	14	5	22	15	11	17

One dot would then be plotted for each examinee, giving a total of only eight dots, yet displaying 16 pieces of numerical information. For example, Candidate 1 studied for seven days and received a score of 23. Candidate 1's dot would be plotted at a vertical of 23 and a horizontal of 7 (see Figure 2-14).

Figure 2-14 A representative point of data on a scatter plot.

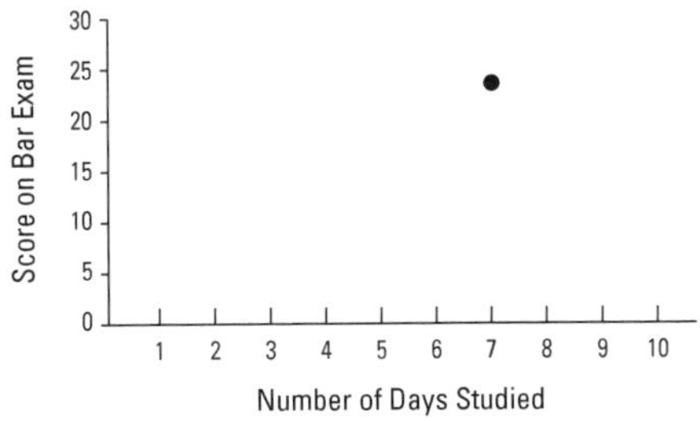

A completed scatter plot would look like Figure 2-15.

Figure 2-15 A scatter plot displaying the relationship between preparation time and test score.

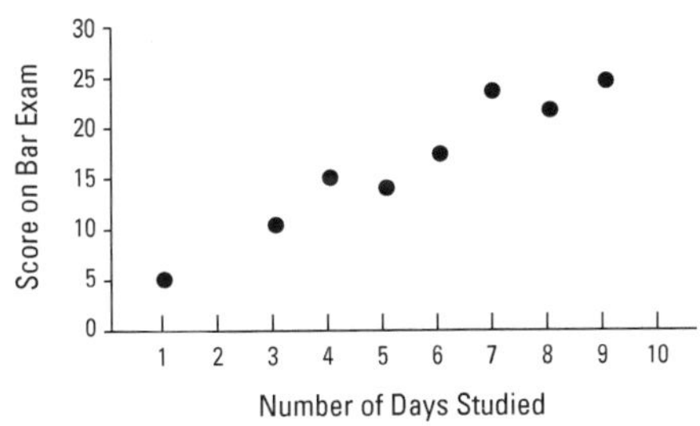

There is a strong **positive relationship** between the number of days studied and the score on the bar exam; that is, the data displayed indicates that an increase in days studied for the exam correlates with an increase in score achieved. A **negative relationship** would be indicated if the dots suggested a line going down from left to right, meaning that as one variable increases, the other decreases. And no relationship would be indicated if the scatter plot dots suggested a completely horizontal line, a completely vertical line, or no line at all (see Figure 2-16).

Figure 2-16 Scatter plots that display no relationship between the variables plotted.

These relationships are discussed in more detail in Chapter 8, which discusses bivariate relationships.

Chapter Check-Out

1. A car salesman takes inventory and finds that he has a total of 125 cars to sell. Of these, 97 are the 2011 model, 11 are the 2010 model, 12 are the 2009 model, and 5 are the 2008 model. Which two types of charts are most appropriate to display the data? Construct one of the plots.
2. Given the bar graph shown here, take the information and construct a scatter plot.

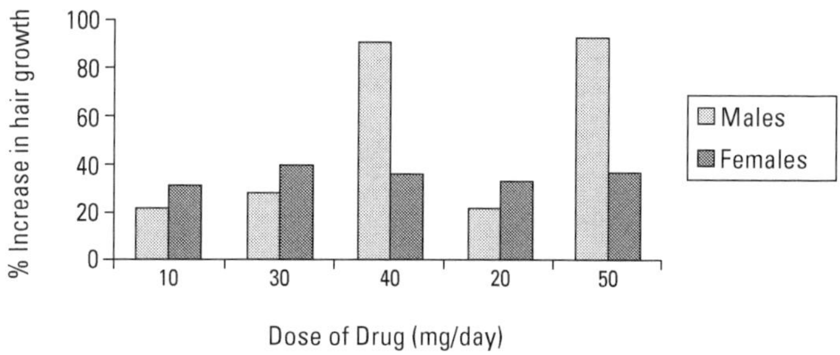

3. Using this ogive, answer the following questions:

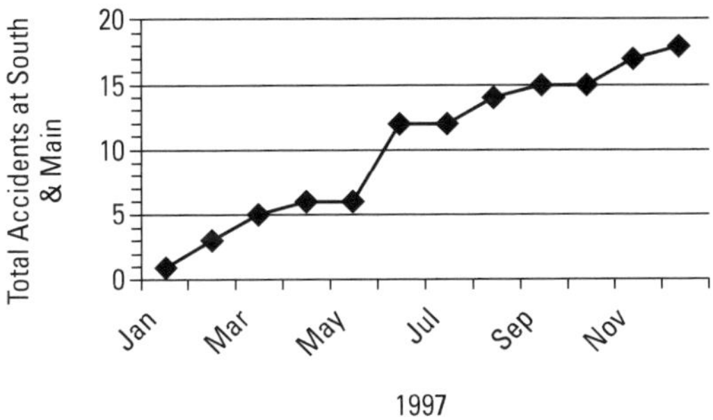

a. How many total accidents have occurred at the intersection of South and Main streets in the last year?

b. How many occurred between June and August?

c. What interval saw the greatest increase in the rate of accidents?

Answers: 1. Pie or bar **2.** Provide your own answer. **3a.** 18 **b.** 2 **c.** May through June

Chapter 3
NUMERICAL MEASURES

Chapter Check-In

- ❑ Calculating and understanding the central tendencies of data
- ❑ Using measures of variability to describe the spread in data
- ❑ Discussing the principal types of measurement scales

Chapter 2 discusses how to graph numerical information—to transform it into a picture—to get a better sense of the data. Another way to get a sense of data is to use numerical measures, certain numbers that give special insight into your values. Two types of numerical measures are important in statistics: **measures of central tendency** and **measures of variation.** Each of these individual measures can provide information about the entire set of data.

Measures of Central Tendency

Measures of central tendency are numbers that tend to cluster around the "middle" of a set of values. Three such middle numbers are the **mean,** the **median,** and the **mode.**

For example, suppose your earnings for the past week were the values shown in Table 3-1.

Table 3-1 Earnings for the Past Week

Day	***Amount***
Monday	\$350
Tuesday	\$150
Wednesday	\$100

(continued)

Table 3-1 *(continued)*

Day	*Amount*
Thursday	\$350
Friday	\$50

Mean

You could express your daily earnings from Table 3-1 in a number of ways. One way is to use the average, or **mean,** of the data set. The arithmetic mean is the sum of the measures in the set divided by the number of measures in the set. Totaling all the measures and dividing by the number of measures, you get \$1,000 ÷ 5 = \$200.

Median

Another measure of central tendency is the **median,** which is defined as the middle value when the numbers are arranged in increasing or decreasing order. When you order the daily earnings shown in Table 3-1, you get \$50, \$100, \$150, \$350, and \$350. The middle value is \$150; therefore, \$150 is the median.

If there is an even number of items in a set, the median is the average of the two middle values. For example, if we had four values—4, 10, 12, and 26—the median would be the average of the two middle values, 10 and 12; in this case, 11 is the median. The median may sometimes be a better indicator of central tendency than the mean, especially when there are **outliers,** or extreme values.

Example 1: Given the four annual salaries of a corporation shown in Table 3-2, determine the mean and the median.

Table 3-2 Four Annual Salaries

Position	*Salary*
CEO	\$1,000,000
Manager	\$50,000
Administrative assistant	\$30,000
Custodian	\$20,000

The mean of these four salaries is \$275,000. The median is the average of the middle two salaries, or \$40,000. In this instance, the median appears to be a better indicator of central tendency because the CEO's salary is an extreme outlier, causing the mean to lie far from the other three salaries.

Mode

Another indicator of central tendency is the **mode,** or the value that occurs most often in a set of numbers. In the set of weekly earnings in Table 3-1, the mode would be $350 because it appears twice and the other values appear only once.

Notation and formulae

The mean of a sample is typically denoted as $\bar{x}$ (read as *x* bar). The mean of a population is typically denoted as μ (pronounced mew). The sum (or total) of measures is typically denoted with a Σ. The formula for a sample mean is

$$\bar{x} = \frac{\sum x}{n} = \frac{x_1 + x_2 + \ldots + x_n}{n}$$

where *n* is the number of values.

Mean for grouped data

Occasionally, you may have data that consist not of actual values but rather of **grouped measures.** For example, you may know that, in a certain working population, 32 percent earn between $25,000 and $29,999; 40 percent earn between $30,000 and $34,999; 27 percent earn between $35,000 and $39,999; and the remaining 1 percent earn between $80,000 and $85,000. This type of information is similar to that presented in a frequency table. (Refer to Chapter 2 for information about frequency tables and graphic displays.) Although you do not have precise individual measures, you still can compute measures for **grouped data,** data presented in a frequency table.

The formula for a sample mean for grouped data is

$$\bar{x} = \frac{\sum fx}{n}$$

where *x* is the midpoint of the interval, *f* is the frequency for the interval, *fx* is the product of the midpoint times the frequency, and *n* is the number of values.

For example, if 8 is the midpoint of a class interval and there are ten measurements in the interval, $fx = 10(8) = 80$, the sum of the ten measurements in the interval.

Σfx denotes the sum of all the products in all class intervals. Dividing that sum by the number of measurements yields the sample mean for grouped data.

For example, consider the information shown in Table 3-3.

Table 3-3 Distribution of the Prices of Items Sold at a Garage Sale

Class Interval	***Frequency (f)***	***Midpoint (x)***	***fx***
\$1.00 to \$5.99	8	3	24
\$6.00 to \$10.99	6	8	48
\$11.00 to \$15.99	4	13	52
\$16.00 to \$20.99	2	18	36
\$21.00 to \$25.99	4	23	92
\$26.00 to \$30.99	6	28	168
\$31.00 to \$35.99	2	33	66
	$n = 32$		$\Sigma fx = 486$

Substituting into the formula:

$$\bar{x} = \frac{\sum fx}{n} = \frac{486}{32} = 15.19$$

Therefore, the average price of items sold was about \$15.19. The value may not be the exact mean for the data, because the actual values are not always known for grouped data.

Median for grouped data

As with the mean, the median for grouped data may not necessarily be computed precisely because the actual values of the measurements may not be known. In that case, you can find the particular interval that contains the median and then approximate the median.

Using Table 3-3, you can see that there is a total of 32 measures. The median is between the 16th and 17th measure; therefore, the median falls within the \$11.00 to \$15.99 interval. The formula for the best approximation of the median for grouped data is

$$\text{median} = L + \frac{w}{f_{\text{med}}}\left(0.5n - \sum f_b\right)$$

where L is the lower class limit of the interval that contains the median, n is the total number of measurements, w is the class width, f_{med} is the frequency of the class containing the median, and Σf_b is the sum of the frequencies for all classes before the median class.

Consider the information in Table 3-4.

Table 3-4 Distribution of Prices of Items Sold at a Garage Sale

Class Boundaries	*Frequency (f)*
$1.00 to $5.99	8
$6.00 to $10.99	6
$11.00 to $15.99	4
$16.00 to $20.99	2
$21.00 to $25.99	4
$26.00 to $30.99	6
$31.00 to $35.99	2
	$n = 32$

As we already know, the median is located in class interval $11.00 to $15.99. So $L = 11$, $n = 32$, $w = 4.99$, $f_{med} = 4$, and $\Sigma f_b = 14$.

Substituting into the formula:

$$
\begin{aligned}
\text{median} &= L + \frac{w}{f_{\text{med}}}\left(0.5n - \sum f_b\right) \\
&= 11 + \frac{4.99}{4}(0.5(32) - 14) \\
&= 11 + \frac{4.99}{4}(16 - 14) \\
&= 11 + \frac{4.99}{4}(2) \\
&= 11 + 2.495 \\
&= 13.495
\end{aligned}
$$

Symmetric distribution

In a distribution displaying perfect symmetry, the mean, the median, and the mode are all at the same point, as shown in Figure 3-1.

Figure 3-1 For a symmetric distribution, mean, median, and mode are equal.

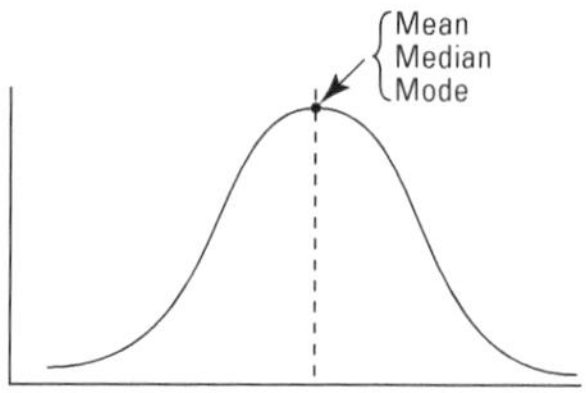

Skewed curves

As you have seen, an outlier can significantly alter the mean of a series of numbers, whereas the median will remain at the center of the series. In such a case, the resulting curve drawn from the values will appear to be **skewed,** tailing off rapidly to the left or right. In the case of negatively skewed or positively skewed curves, the median remains in the center of these three measures.

Figure 3-2 shows a negatively skewed curve.

Figure 3-2 A negatively skewed distribution, mean < median < mode.

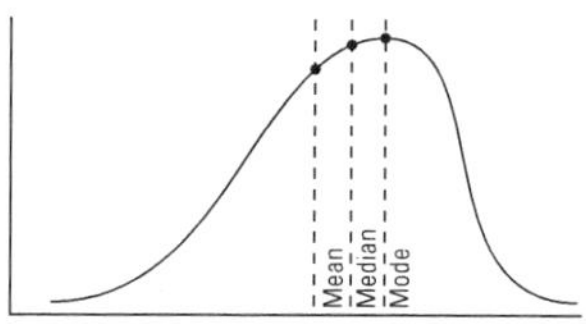

Figure 3-3 shows a positively skewed curve.

Figure 3-3 A positively skewed distribution, mode < median < mean.

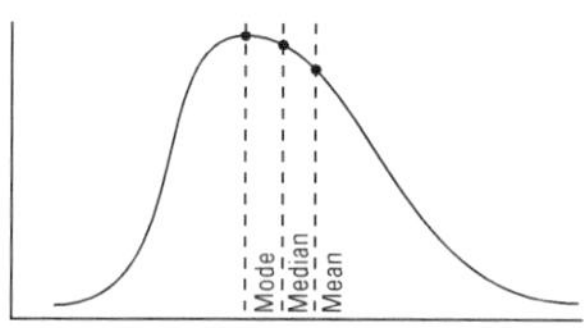

Measures of Variability

Measures of central tendency locate only the center of a distribution of measures. Other measures often are needed to describe data.

For example, consider the two sets of numbers presented in Table 3-5.

Table 3-5 Earnings of Two Employees

Daily Earnings of Employee A	*Daily Earnings of Employee B*
\$200	\$200
\$210	\$20
\$190	\$400
\$201	\$0
\$199	\$390
\$195	\$10
\$205	\$200
\$200	\$380

The mean, the median, and the mode of each employee's daily earnings all equal \$200. Yet, there is significant difference between the two sets of numbers. For example, the daily earnings of Employee A are much more consistent than those of Employee B, which show great variation. This example illustrates the need for **measures of variation** or spread.

Range

The most elementary measure of variation is **range.** Range is defined as the difference between the largest and smallest values. The range is a single number. To say that the range is from \$190 to \$200, although informative, is not really a correct use of the term. The range for Employee A is \$210 – \$190 = \$20; the range for Employee B is \$400 – \$0 = \$400.

Deviation and variance

The deviation is defined as the distance of the measurements away from the mean. In Table 3-5, Employee A's earnings have considerably less deviation than Employee B's do. The **variance** is defined as the sum of the squared deviations of n measurements from their mean divided by $(n - 1)$.

So, from Table 3-5, the mean for Employee A is $200, and the deviations from the mean are as follows:

0, +10, –10, +1, –1, –5, +5, 0

The squared deviations from the mean, therefore, are the following:

0, 100, 100, 1, 1, 25, 25, 0

The sum of these squared deviations from the mean equals 252. Dividing by (n – 1), or 8 – 1, yields $\frac{252}{7} = 36$. So, the variance is 36.

For Employee B, the mean is also $200, and the deviations from the mean are as follows:

0, –180, +200, –200, +190, –190, 0, +180

The squared deviations, therefore, are the following:

0; 32,400; 40,000; 40,000; 36,100; 36,100; 0; 32,400

The sum of these squared deviations equals 217,000. Dividing by (n – 1) yields $\frac{217{,}000}{7} = 31{,}000$.

Although they earned the same totals, there is significant difference in variance between the daily earnings of the two employees.

Standard deviation

The **standard deviation** is defined as the positive square root of the variance; thus, the standard deviation of Employee A's daily earnings is the positive square root of 36, or 6. The standard deviation of Employee B's daily earnings is the positive square root of 31,000, or about 176.

Notation

s^2 denotes the variance of a sample.

σ^2 denotes the variance of a population.

s denotes the standard deviation of a sample.

σ denotes the standard deviation of a population.

Empirical rule: The normal curve

One practical significance of the standard deviation is that, with mound-shaped (bell-shaped) distributions, the following rules apply:

- The interval from one standard deviation below the mean to one standard deviation above the mean contains approximately 68 percent of the measurements.
- The interval from two standard deviations below the mean to two standard deviations above the mean contains approximately 95 percent of the measurements.
- The interval from three standard deviations below the mean to three standard deviations above the mean contains approximately all the measurements.

These mound-shaped curves usually are called **normal distributions** or normal curves (see Figures 3-4, 3-5, and 3-6).

Figure 3-4 The interval $\pm\sigma$ from the mean contains 68 percent of the measurements.

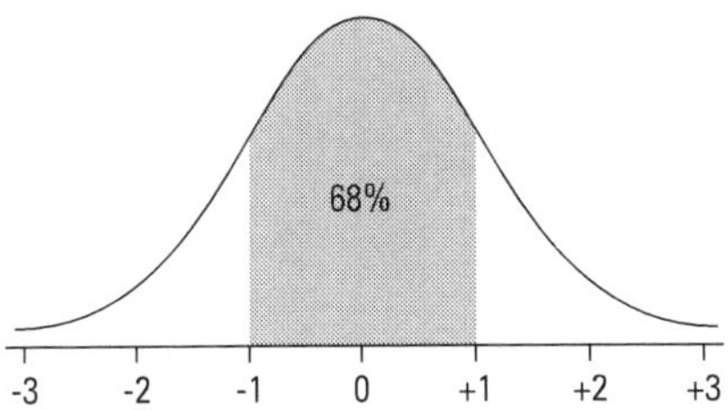

Figure 3-5 The interval $\pm 2\sigma$ from the mean contains 95 percent of the measurements.

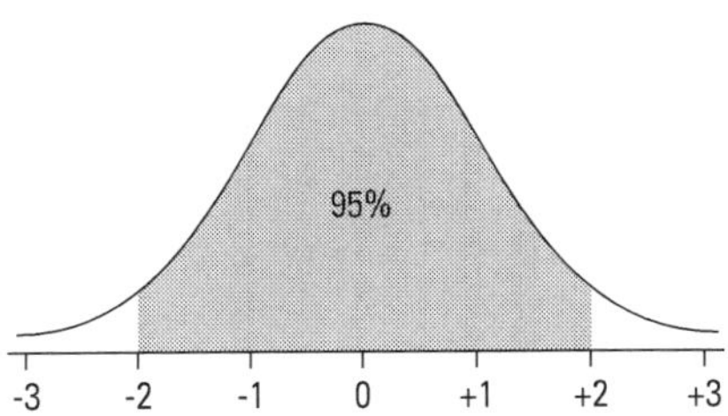

Figure 3-6 The interval ±3σ from the mean contains 99.7 percent of the measurements.

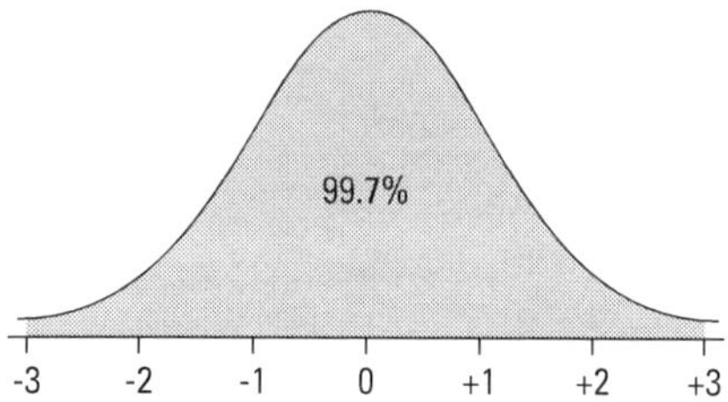

Shortcut formulae

A shortcut method of calculating variance and standard deviation requires two quantities: sum of the values and sum of the squares of the values.

Σx = sum of the measures

Σx^2 = sum of the squares of the measures

For example, using these six measures: 3, 9, 1, 2, 5, and 4:

$$\Sigma x = 3 + 9 + 1 + 2 + 5 + 4 = 24$$

$$\Sigma x^2 = 3^2 + 9^2 + 1^2 + 2^2 + 5^2 + 4^2$$

$$= 9 + 81 + 1 + 4 + 25 + 16 = 136$$

The quantities are then substituted into the shortcut formula to find $\sum(x-\bar{x})^2$.

$$\sum(x-\bar{x})^2 = \sum x^2 - \frac{\left(\sum x\right)^2}{n}$$

$$= 136 - \frac{(24)^2}{6}$$

$$= 136 - \frac{576}{6}$$

$$= 40$$

The variance and standard deviation are now found as before:

$$s^2 = \frac{\sum(x-\bar{x})^2}{n-1} = \frac{40}{5} = 8$$

$$s = \sqrt{s^2} = \sqrt{8} = 2.828$$

Percentile

The *N*th **percentile** is defined as the value such that *N* percent of the values lie below it. So, for example, a score of 5 percent from the top score would be the 95th percentile because it is above 95 percent of the other scores (see Figure 3-7).

Figure 3-7 Ninety-five percent of the test scores are less than the value of the 95th percentile.

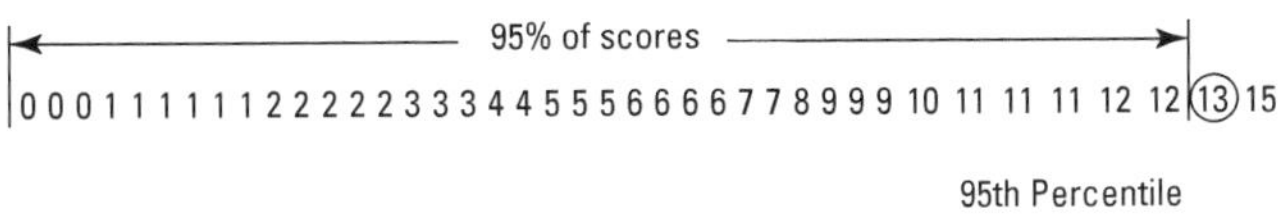

Quartiles and interquartile range

The **lower quartile** (Q_1) is defined as the 25th percentile; thus, 75 percent of the measures are above the lower quartile. The **middle quartile** (Q_2) is defined as the 50th percentile, which is, in fact, the median of all the measures. The **upper quartile** (Q_3) is defined as the 75th percentile; thus, only 25 percent of the measures are above the upper quartile.

The **interquartile range** (*IQR*) is the value for $Q_3 - Q_1$. Like the range, it is a single value.

Figure 3-8 illustrates the locations of the median and the quartiles for a set of 20 test scores.

Figure 3-8 An illustration of the quartiles and the interquartile range.

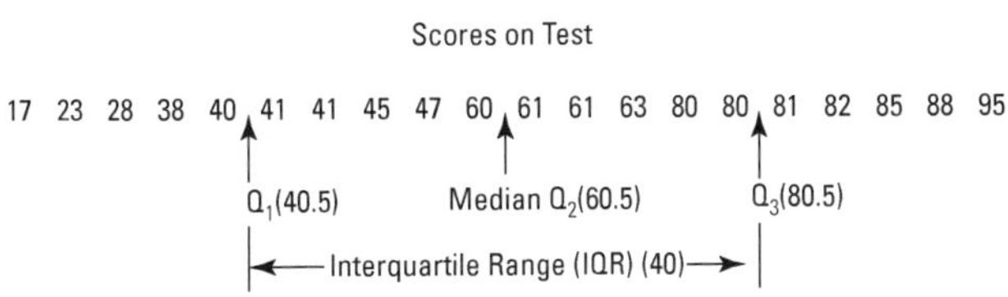

Measurement Scales

Different measurement scales allow for different levels of exactness, depending upon the characteristics of the variables being measured. The four types of scales available in statistical analysis are

- **Nominal:** A scale that measures data by name only. For example, religious affiliation (measured as Christian, Jewish, Muslim, and so forth), political affiliation (measured as Democratic, Republican, Libertarian, and so forth), or style of automobile (measured as sedan, sports car, SUV, and so forth).
- **Ordinal:** A scale that measures by rank order only. Other than rough order, no precise measurement is possible. For example, medical condition (measured as satisfactory, fair, poor, guarded, serious, and critical); socioeconomic status (measured as lower class, lower-middle class, middle class, upper-middle class, upper class); or military officer rank (measured as lieutenant, captain, major, lieutenant colonel, colonel, general). Such rankings are not absolute but rather relative to each other: Major is higher than captain, but we cannot measure the exact difference in numerical terms. Is the difference between major and captain equal to the difference between colonel and general? We cannot say.
- **Interval:** A scale that measures by using equal intervals. Here you can compare differences between pairs of values. The Fahrenheit temperature scale, measured in degrees, is an interval scale, as is the centigrade scale. The temperature difference between 50°C and 60°C (10 degrees) equals the temperature difference between 80°C and 90°C (10 degrees). Note that the 0 in each of these scales is arbitrarily placed, which makes the interval scale different from ratio.
- **Ratio:** Similar to an interval scale, a ratio scale includes a 0 measurement that signifies the point at which the characteristic being measured vanishes (absolute zero). For example, income (measured in dollars, with 0 equal to no income at all), years of formal education, items sold, and so forth, are all ratio scales.

Chapter Check-Out

Q&A

1. A doodad machine is capable of making up to 200 doodads per hour.

a. For four consecutive hours, the doodad machine makes 120, 118, 124, and 112 doodads. Find the average rate of doodad production.

b. During the fifth hour, the machine breaks after making only 21 doodads. The remainder of the hour is spent repairing it. Normally, the machine must average 115 doodads per hour, over the course of an eight-hour day, in order to meet quota. What must the average rate be for the next three hours to meet this quota?

2. A meteorologist measures the average rate of rainfall in order to test the saying "When it rains, it pours." Her data, showing the rainfall rate over the course of a rainy morning, are as follows:

Time of Day	*Rainfall Rate (inches/minute)*
10:00 a.m. to 10:15 a.m.	0.011
10:15 a.m. to 10:30 a.m.	0.007
10:30 a.m. to 10:45 a.m.	0.001
10:45 a.m. to 11:00 a.m.	0.010
11:00 a.m. to 11:15 a.m.	0.003
11:15 a.m. to 11:30 a.m.	0.045
11:30 a.m. to 11:45 a.m.	0.002
11:45 a.m. to 12:00 p.m.	0.006

a. Find the median and mean.
b. Calculate the variance and standard deviation.

3. Can a distribution have the mean equal to the median, but the mode not equal to either mean or median? If so, draw it.

Answers: 1a. 118.5 **b.** 141.67 **2a.** 0.0065, 0.0106 **b.** 0.0002, 0.014 **3.** Yes (a symmetric but bimodal distribution is a good example).

Chapter 4
PROBABILITY

Chapter Check-In

- ❑ Applying classic probability theory to simple events
- ❑ Analyzing combinations of simple events based on these rules
- ❑ Learning about probability distributions, especially the binomial distribution

Probability theory plays a central role in statistics. After all, statistical analysis is applied to a collection of data in order to discover something about the underlying events. These events may be connected to one another—for example, mutually exclusive—but the individual choices involved are assumed to be random. Alternatively, we may sample a population at random and make inferences about the population as a whole from the sample by using statistical analysis. Therefore, a solid understanding of probability theory—the study of random events—is necessary to understand how statistical analysis works and also to correctly interpret the results.

You have an intuition about probability. As you will see, in some cases, probability theory seems obvious. But be careful: Occasionally, a seemingly obvious answer will turn out to be wrong—because sometimes your intuition about probability will fail. Even in seemingly simple cases, it is best to follow the rules of probability as described in this chapter rather than rely on your hunches.

Classic Theory

The classic theory of probability underlies much of probability in statistics. Briefly, this theory states that the chance of a particular outcome occurring is determined by the ratio of the number of favorable outcomes (or "successes") to the total number of outcomes. Expressed as a formula,

$$P(A) = \frac{\text{number of favorable outcomes}}{\text{total number of possible outcomes}}$$

For example, the probability of randomly drawing an ace from a well-shuffled deck of cards is equal to the ratio $\frac{4}{52}$. Four is the number of favorable outcomes (the number of aces in the deck), and 52 is the number of total outcomes (the number of cards in the deck). The probability of randomly selecting an ace in one draw from a deck of cards is, therefore, $\frac{4}{52}$, or 0.077. In statistical analysis, probability is usually expressed as a decimal and ranges from a low of 0 (no chance) to a high of 1 (certainty).

The classic theory assumes that all outcomes have equal likelihood of occurring. In the example just cited, each card must have an equal chance of being chosen—no card is larger than any other or in any way more likely to be chosen than any other card.

The classic theory pertains only to outcomes that are **mutually exclusive** (or **disjoint**), which means that those outcomes may not occur at the same time. For example, one coin flip can result in a head or a tail, but one coin flip cannot result in a head *and* a tail. So, the outcome of a head and the outcome of a tail are said to be "mutually exclusive" in one coin flip, as is the outcome of an ace and a king in one card being drawn.

Relative Frequency Theory

The **relative frequency theory of probability** holds that if an experiment is repeated an extremely large number of times and a particular outcome occurs a percentage of the time, then that particular percentage is close to the probability of that outcome.

For example, if a machine produces 10,000 widgets one at a time, and 1,000 of those widgets are faulty, the probability of that machine producing a faulty widget is approximately 1,000 out of 10,000, or 0.10.

Probability of Simple Events

Example 1: What is the probability of simultaneously flipping three coins—a penny, a nickel, and a dime—and having all three land heads?

Using the classic theory, determine the ratio of the number of favorable outcomes to the number of total outcomes. Table 4-1 lists all possible outcomes.

Table 4-1 Possible Outcomes of Penny, Nickel, and Dime Flipping

Outcome	*Penny*	*Nickel*	*Dime*
1	Head	Head	Head
2	Head	Head	Tail
3	Head	Tail	Head
4	Head	Tail	Tail
5	Tail	Head	Head
6	Tail	Head	Tail
7	Tail	Tail	Head
8	Tail	Tail	Tail

There are eight different outcomes, only one of which is favorable (Outcome 1, all three coins landing heads); therefore, the probability of three coins landing heads is $\frac{1}{8}$, or 0.125.

What is the probability of exactly two of the three coins landing heads? Again, there are the eight total outcomes, but in this case only three favorable outcomes (outcomes 2, 3, and 5); thus, the probability of exactly two of three coins landing heads is $\frac{3}{8}$, or 0.375.

Independent Events

Each of the three coins being flipped in the preceding example is what is known as an **independent event.** Independent events are defined as outcomes that are not affected by other outcomes. In other words, the flip of the penny does not affect the flip of the nickel, and vice versa.

Dependent Events

Dependent events, on the other hand, are outcomes that are affected by other outcomes. Consider the following example.

Example 2: What is the probability of randomly drawing an ace from a deck of cards and then drawing an ace again from the same deck of cards, without returning the first drawn card back to the deck?

For the first draw, the probability of a favorable outcome is $\frac{4}{52}$, as explained earlier; however, after that first card has been drawn, the total number of outcomes is no longer 52, but now 51 because a card has been removed from the deck. And if that first card drawn resulted in a favorable outcome (an ace), there would now be only three aces in the deck. If that first card drawn were *not* an ace, the number of favorable outcomes would remain at four. So, the second draw is a dependent event because its probability changes depending upon what happens on the first draw.

If, however, you replace that drawn card back into the deck and shuffle well again before the second draw, then the probability for a favorable outcome for each draw will now be equal $\left(\frac{4}{52}\right)$, and these events will be independent.

Probability of Joint Occurrences

Another way to compute the probability of all three flipped coins landing heads is as a series of three different events: First flip the penny, then flip the nickel, and then flip the dime. Will the probability of landing three heads still be 0.125?

Multiplication rule

To compute the probability of **joint occurrence** (two or more independent events all occurring), multiply their probabilities.

For example, the probability of the penny landing heads is $\frac{1}{2}$, or 0.5; the probability of the nickel next landing heads is $\frac{1}{2}$, or 0.5; and the probability of the dime landing heads is $\frac{1}{2}$, or 0.5. Thus, note that

$$0.5 \times 0.5 \times 0.5 = 0.125$$

which is what you determined with the classic theory by assessing the ratio of the number of favorable outcomes to the number of total outcomes. The notation for joint occurrence is

$$P(A \cap B) = P(A) \times P(B)$$

which is read: The probability of A and B both happening is equal to the probability of A times the probability of B.

Using the **multiplication rule,** you also can determine the probability of drawing two aces in a row from a deck of cards. The only way to draw two aces in a row from a deck of cards is for both draws to be favorable. For the first draw, the probability of a favorable outcome is $\frac{4}{52}$. But because the first draw is favorable, only three aces are left among 51 cards. So, the probability of a favorable outcome on the second draw is $\frac{3}{51}$. For both events to happen, you simply multiply those two probabilities together:

$$\frac{4}{52} \times \frac{3}{51} = \frac{12}{2{,}652} = 0.0045$$

Note that these probabilities are not independent. If, however, you had decided to return the initial card drawn back to the deck before the second draw, then the probability of drawing an ace on each draw is $\frac{4}{52}$, because these events are now independent. Drawing an ace twice in a row, with the odds being $\frac{4}{52}$ both times, gives the following:

$$\frac{4}{52} \times \frac{4}{52} = \frac{16}{2{,}704} = 0.0059$$

In either case, you use the multiplication rule because you are computing probability for favorable outcomes in all events.

Addition rule

Given mutually exclusive events, finding the probability of *at least one* of them occurring is accomplished by adding their probabilities.

For example, what is the probability of one coin flip resulting in at least one head or at least one tail?

The probability of one coin flip landing heads is 0.5, and the probability of one coin flip landing tails is 0.5. Are these two outcomes mutually exclusive in one coin flip? Yes, they are. You cannot have a coin land both heads and tails in one coin flip; therefore, you can determine the probability of at least one head or one tail resulting from one flip by adding the two probabilities:

$$0.5 + 0.5 = 1 \text{ (or certainty)}$$

Example 3: What is the probability of at least one spade or one club being randomly chosen in one draw from a deck of cards?

The probability of drawing a spade in one draw is $\frac{13}{52}$; the probability of drawing a club in one draw is $\frac{13}{52}$. These two outcomes are mutually exclusive in one draw because you cannot draw both a spade and a club in one draw; therefore, you can use the **addition rule** to determine the probability of drawing at least one spade or one club in one draw:

$$\frac{13}{52}+\frac{13}{52}=\frac{26}{52}=0.50$$

Non-Mutually-Exclusive Outcomes

For the addition rule to apply, the events must be mutually exclusive. Now consider the following example.

Example 4: What is the probability of the outcome of at least one head in two coin flips?

Should you add the two probabilities as in the preceding examples? In the first example, you added the probability of getting a head and the probability of getting a tail because those two events were mutually exclusive in one flip. In the second example, the probability of getting a spade was added to the probability of getting a club because those two outcomes were mutually exclusive in one draw. Now when you have two flips, should you add the probability of getting a head on the first flip to the probability of getting a head on the second flip? Are these two events mutually exclusive?

Of course, they are not mutually exclusive. You can get an outcome of a head on one flip and a head on the second flip. So, because they are not mutually exclusive, you cannot use the addition rule. If you did use the addition rule, you would get

$$\frac{1}{2}+\frac{1}{2}=\frac{2}{2}=1$$

or certainty, which is absurd. There is no certainty of getting at least one head on two flips. (Try it several times, and see that there is a possibility of getting two tails and no heads.)

Double-Counting

By using the addition rule in a situation that is not mutually exclusive, you are **double-counting.** One way of realizing that you are double-counting is to use the classic theory of probability: List all the different outcomes when flipping a coin twice and assess the ratio of favorable outcomes to total outcomes (see Table 4-2).

Table 4-2 All Possible Outcomes of Flipping the Same Coin Twice

First Flip	*with*	*Second Flip*
Head	+	Head
Head	+	Tail
Tail	+	Head
Tail	+	Tail

There are four total outcomes. Three of the outcomes have at least one head; therefore, the probability of throwing at least one head in two flips is $\frac{3}{4}$, or 0.75, not 1. But if you had used the addition rule, you would have added the two heads from the first flip to the two heads from the second flip and gotten four heads in four flips, $\frac{4}{4} = 1$. But the two heads in that first pair constitute only one outcome; so, by counting both heads for that outcome, you are double-counting because this is the joint-occurrence outcome that is not mutually exclusive.

To use the addition rule in a non-mutually-exclusive situation, you must subtract any events that double-count. In this case:

$$\frac{1}{2} + \frac{1}{2} - \frac{1}{4} = \frac{3}{4} = 0.75$$

The notation, therefore, for at least one favorable occurrence in two events is

$$P(A \cup B) = P(A) + P(B) - P(A \cap B)$$

which is read: The probability of at least one of the events *A* or *B* equals the probability of *A* plus the probability of *B* minus the probability of their joint occurrence. (Note that if they are mutually exclusive, then $P(A \cap B)$—the joint occurrence—equals 0, and you simply add the two probabilities.)

Example 5: What is the probability of drawing either a spade or an ace from a deck of cards?

The probability of drawing a spade is $\frac{13}{52}$; the probability of drawing an ace is $\frac{4}{52}$. But the probability of their joint occurrence (an ace of spades) is $\frac{1}{52}$. Thus,

$$\begin{aligned} P(A \cup B) &= P(A) + P(B) - P(A \cap B) \\ &= \frac{13}{52} + \frac{4}{52} - \frac{1}{52} \\ &= \frac{16}{52} \\ &= \frac{4}{13} \\ &= 0.3077 \end{aligned}$$

Conditional Probability

Sometimes you have more information than simply total outcomes and favorable outcomes; hence, you are able to make more informed judgments regarding probabilities. For example, suppose you know the following information: In a particular village, there are 60 women and 40 men. Twenty of those women are 70 years of age or older; five of the men are 70 years of age or older (see Table 4-3).

Table 4-3 Distribution of People in a Particular Village

	70 or older	*69 or younger*	*Totals*
Women	20	40	60
Men	5	35	40
Total	25	75	100

What is the probability that a person selected at random in that town will be a woman? Because women constitute 60 percent of the total population, the probability is 0.60.

What is the probability that a person 70 years of age or older, selected at random, will be a woman? This question is different because the probability of A (being a woman) given B (the person in question being 70 years of age or older) is now conditional upon B (being 70 years of age or older).

Because women number 20 out of the 25 people in the 70-or-older group, the probability of this latter question is $\frac{20}{25}$, or 0.80.

Conditional probability is found using this formula:

$$P(A|B)=\frac{P(A\cap B)}{P(B)}$$

which is read: The probability of *A* given *B* equals the probability of *A* and *B* divided by the probability of *B*. The vertical bar in the expression $A|B$ is read *given that* or *given*.

Probability Distributions

A **probability distribution** is a pictorial display of the probability—$P(x)$ —for any value of *x*. Consider the number of possible outcomes of two coins being flipped (see Table 4-4). Table 4-5 shows the probability distribution of the results of flipping two coins. Figure 4-1 displays this information graphically.

Table 4-4 Possible Outcomes of Two Flipped Coins

Head + head	=	2 heads
Head + tail	=	1 head
Tail + head	=	1 head
Tail + tail	=	0 heads

Table 4-5 Probability Distribution: Number of Heads

x	*P(x)*
0	$\frac{1}{4}$, or 0.25
1	$\frac{1}{2}$, or 0.50
2	$\frac{1}{4}$, or 0.25

Figure 4-1 Probability distribution of the results of flipping two coins.

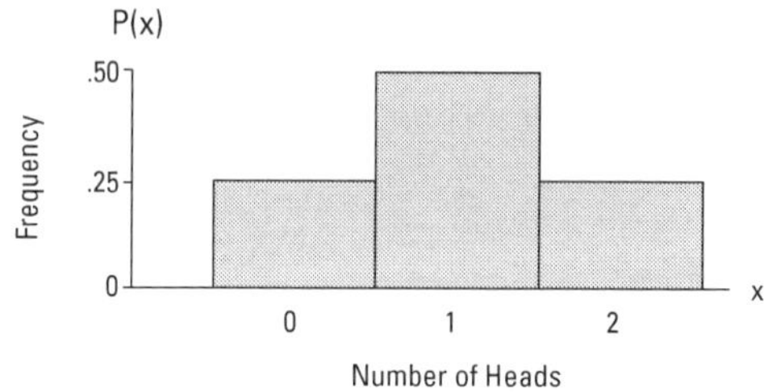

Discrete versus continuous variables

The number of heads resulting from coin flips can be counted only in integers (whole numbers). The number of aces drawn from a deck can be counted only in integers. These "countable" numbers are known as **discrete variables:** 0, 1, 2, 3, 4, and so forth. Nothing in between two variables is possible. For example, 2.6 is not possible.

Qualities such as height, weight, temperature, distance, and so forth, however, can be measured using fractions or decimals as well: 34.27, 102.26, and so forth. These are known as **continuous variables.**

Total of probabilities of discrete variables

The probability of a discrete variable lies somewhere between 0 and 1, inclusive. For example, the probability of tossing one head in two coin flips is, as earlier, 0.50. The probability of tossing two heads in two coin flips is 0.25, and the probability of tossing no heads in two coin flips is 0.25. The sum (total) of probabilities for all values of x always equals 1. For example, note that in Tables 4-3 and 4-5, adding the probabilities for all the possible outcomes yields a sum of 1.

The Binomial

A discrete variable that can result in only one of two outcomes is called **binomial.** For example, a coin flip is a binomial variable, but drawing a card from a standard deck of 52 is not. Whether a drug is successful or unsuccessful in producing results is a binomial variable, as is whether a machine produces perfect or imperfect widgets.

Binomial experiments

Binomial experiments require the following elements:

- The experiment consists of a number of identical events (n).
- Each event has only one of two mutually exclusive outcomes. (These outcomes are called successes and failures.)
- The probability of a success outcome is equal to some percentage, which is identified as a **proportion, π**.
- This proportion, π, remains constant throughout all events and is defined as the ratio of number of successes to number of trials.
- The events are independent.
- Given all the above, the **binomial** formula can be applied (x = number of favorable outcomes; n = number of events):

$$P(x) = \frac{n!}{x!(n-x)!}\pi^x(1-\pi)^{n-x}$$

$$n! = n(n-1)(n-2)\ldots(3)(2)(1)$$

Example 6: A coin is flipped ten times. What is the probability of getting exactly five heads? Using the binomial formula, where n (the number of events) is given as 10; x (the number of favorable outcomes) is given as 5; and the probability of landing a head in one flip is 0.5:

$$\begin{aligned} P(x) &= \frac{n!}{x!(n-x)!}\pi^x(1-\pi)^{n-x} \\ &= \frac{10!}{5!5!}(0.5^5)(1-0.5)^5 \\ &= \frac{10\cdot9\cdot8\cdot7\cdot6\cdot5\cdot4\cdot3\cdot2\cdot1}{5\cdot4\cdot3\cdot2\cdot1\cdot5\cdot4\cdot3\cdot2\cdot1}(0.0313)(0.0313) \\ &= 252(0.0313)(0.0313) \\ &= 0.246 \end{aligned}$$

So, the probability of getting exactly five heads in ten flips is 0.246, or approximately 25 percent.

Binomial table

Because probabilities of binomial variables are so common in statistics, tables are used to alleviate having to continually use the formula. Refer to Table 1 in Appendix B, and you will find that given $n = 10$, $x = 5$, and $\pi = 0.5$, the probability is 0.2461.

Mean and standard deviation

The mean of the binomial probability distribution is determined by the following formula:

$$\mu = n\pi$$

where π is the proportion of favorable outcomes and n is the number of events.

The standard deviation of the binomial probability distribution is determined by this formula:

$$\sigma = \sqrt{n\pi(1-\pi)}$$

Example 7: What is the mean and standard deviation for a binomial probability distribution for ten coin flips of a fair coin?

Because the proportion of favorable outcomes of a fair coin falling heads (or tails) is $\pi = 0.5$, simply substitute into the formulas:

$$\mu = n\pi = 10(0.5) = 5$$

$$\sigma = \sqrt{n\pi(1-\pi)} = \sqrt{10(0.5)(0.5)} = \sqrt{2.5} = 1.581$$

The probability distribution for the number of favorable outcomes is shown in Figure 4-2.

Note that this distribution appears to display symmetry. Only a binomial distribution with $\pi = 0.5$ will be truly symmetric. All other binomial distribution will be skewed. Binomial probability distributions and use of the normal approximation to replace the probability histogram are discussed in Chapter 5.

Figure 4-2 The binomial probability distribution of the number of heads resulting from ten coin tosses.

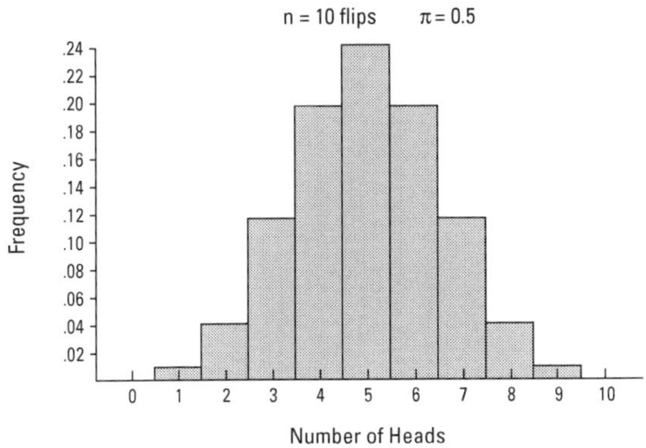

Chapter Check-Out

Q&A

1. A dartboard is divided into 20 equal wedges, ignoring the bull's-eye.
   - **a.** If only 6 of the 20 wedges are worth points, what is the probability of scoring on one throw (assuming you can hit the board)?
   - **b.** What is the chance of hitting at least one scoring region in three consecutive throws?

2. During a boring lecture in statistics class, you begin repeatedly flipping an old, worn quarter. Amazingly, the first ten flips are all heads.
   - **a.** What is the chance of this, if the coin is fair?
   - **b.** Suppose that the coin is not fair. If the probability of heads is 75 percent rather than 50 percent, what is the chance of your result?
   - **c.** How unfair would the coin have to be in order for your result to have had a 50 percent chance of occurring?

3. A woman has three sons. What is the probability that her next child will be a girl?

4. Consider an ordinary (six-sided) die, assumed to be fair.

a. Use the binomial distribution to determine the probability that, in ten rolls of the die, you will see exactly four sixes.

b. If you roll the die 100 times, what is the chance of rolling exactly 40 sixes?

Answers: 1a. 0.3 **b.** 0.657 **2a.** 9.77×10^{-4} **b.** 0.0563 **c.** $P(\text{heads}) = 0.933$ **3.** 0.5 **4a.** 0.0543 **b.** 1.82×10^{-8}

Chapter 5
SAMPLING

Chapter Check-In

- ❑ Examining the differences between populations and samples
- ❑ Learning about sampling distributions, sampling errors, and the central limit theorem
- ❑ Discussing in detail the properties of the normal distribution

As discussed in Chapter 1, the field of inferential statistics enables you to make educated guesses about the numerical characteristics of large groups. The logic of sampling gives you a way to test conclusions about such groups using only a small portion of its members.

Populations, Samples, Parameters, and Statistics

A **population** is a group of phenomena that have something in common. The term often refers to a group of people, as in the following examples:

- All registered voters in Crawford County
- All members of the International Machinists Union
- All Americans who played golf at least once in the past year

But populations can refer to things as well as people:

- All widgets produced last Tuesday by the Acme Widget Company
- All daily maximum temperatures in July for major U.S. cities
- All basal ganglia cells from a particular rhesus monkey

Often, researchers want to know things about populations but do not have data for every person or thing in the population. If a company's customer service division wanted to learn whether its customers were satisfied, it would not be practical (or perhaps even possible) to contact every individual who purchased a product. Instead, the company might select a **sample** of the population. A sample is a smaller group of members of a population selected to represent the population. In order to use statistics to learn things about the population, the sample must be **random.** A random sample is one in which every member of a population has an equal chance of being selected. The most commonly used sample is a simple random sample. It requires that every possible sample of the selected size has an equal chance of being used.

A **parameter** is a characteristic of a population. A **statistic** is a characteristic of a sample. Inferential statistics enables you to make an educated guess about a population parameter based on a statistic computed from a sample randomly drawn from that population (see Figure 5-1).

Figure 5-1 Illustration of the relationship between samples and populations.

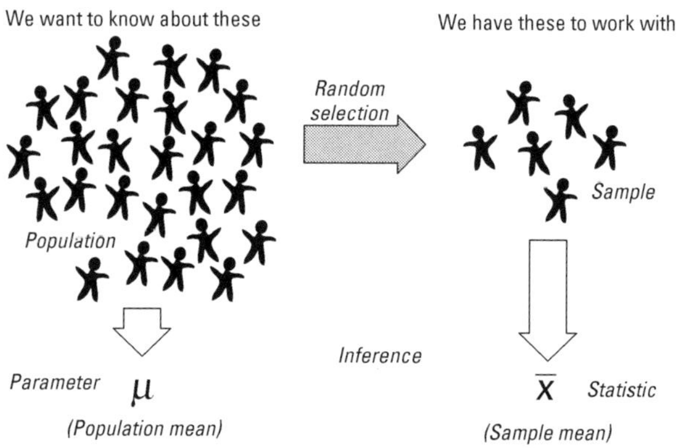

For example, say you want to know the mean income of the subscribers to a particular magazine—a parameter of a population. You draw a random sample of 100 subscribers and determine that their mean income is $27,500 (a statistic). You conclude that the population mean income μ is likely to be close to $27,500 as well. This example is one of statistical inference.

Different symbols are used to denote statistics and parameters, as Table 5-1 shows.

Table 5-1 Comparison of Sample Statistics and Population Parameters

	Sample Statistic	*Population Parameter*
Mean	$\overline{X}$	μ
Standard deviation	s	σ
Variance	s^2	σ^2

Sampling Distributions

Continuing with the earlier example, suppose that ten different samples of 100 people were drawn from the population, instead of just one. You would not expect the income means of these ten samples to be exactly the same, because of **sampling variability** (the tendency of the same statistic computed from a number of random samples drawn from the same population to differ).

Suppose that the first sample of 100 magazine subscribers was "returned" to the population (made available to be selected again), another sample of 100 subscribers was selected at random, and the mean income of the new sample was computed. If this process were repeated ten times, it might yield the following sample means:

27,500	27,192	28,736	26,454	28,527
28,407	27,592	27,684	28,827	27,809

These ten values are part of a **sampling distribution.** The sampling distribution of a statistic (in this case, of a mean) is the distribution obtained by computing the statistic for all possible samples of a specific size drawn from the same population.

You can estimate the mean of this sampling distribution by summing the ten sample means and dividing by ten, which gives a distribution mean of 27,872.8. Suppose that the mean income of the entire population of subscribers to the magazine is $28,000. (You usually do not know what it is.) You can see in Figure 5-2 that the first sample mean ($27,500) was not a bad estimate of the population mean and that the mean of the distribution of ten sample means ($27,872) was even better.

Figure 5-2 Estimation of the population mean becomes progressively more accurate as more samples are taken.

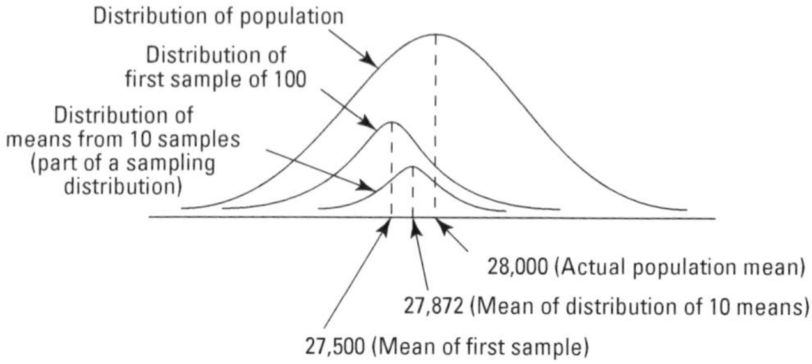

Random and Systematic Error

Two potential sources of error occur in statistical estimation—two reasons a statistic might misrepresent a parameter. **Random error** occurs as a result of sampling variability. The ten sample means in the preceding section differed from the true population mean because of random error. Some were below the true value; some above it. Similarly, the mean of the distribution of ten sample means was slightly lower than the true population mean. If ten more samples of 100 subscribers were drawn, the mean of that distribution—that is, the mean of those means—might be higher than the population mean.

Systematic error or **bias** refers to the tendency to consistently underestimate or overestimate a true value. Suppose that your list of magazine subscribers was obtained through a database of information about air travelers. The samples that you would draw from such a list would likely overestimate the population mean of all subscribers' income because lower-income subscribers are less likely to travel by air and many of them would be unavailable to be selected for the samples. This example would be one of bias.

In Figure 5-3, both of the dot plots on the right illustrate systematic error (bias). The results from the samples for these two situations do not have a center close to the true population value. Both of the dot plots on the left have centers close to the true population value.

Figure 5-3 Random (sampling) error and systematic error (bias) distort the estimation of population parameters from sample statistics.

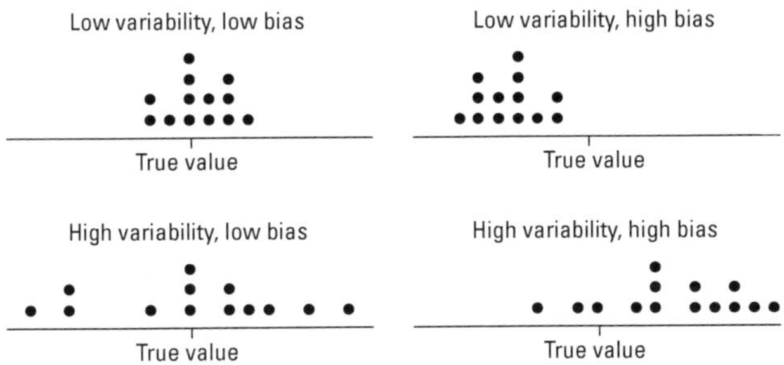

Central Limit Theorem

If the population of all subscribers to the magazine were normal, you would expect its sampling distribution of means to be normal as well. But what if the population were non-normal? The **central limit theorem** states that even if a population distribution is strongly non-normal, its sampling distribution of means will be approximately normal for large sample sizes (over 30). The central limit theorem makes it possible to use probabilities associated with the normal curve to answer questions about the means of sufficiently large samples.

According to the central limit theorem, the mean of a sampling distribution of means is an unbiased estimator of the population mean.

$$\mu_{\bar{x}} = \mu$$

Similarly, the standard deviation of a sampling distribution of means is

$$\sigma_{\bar{x}} = \frac{\sigma}{\sqrt{n}}$$

Note that the larger the sample, the less variable the sample mean. The mean of many observations is less variable than the mean of few. The standard deviation of a sampling distribution of means is often called the **standard error** of the mean. Every statistic has a standard error, which is a measure of the statistic's random variability.

Example 1: If the population mean of number of fish caught per trip to a particular fishing hole is 3.2 and the population standard deviation is 1.8, what are the mean and standard deviation of the sampling distribution for samples of size 40 trips?

$$\mu_{\bar{x}} = 3.2$$

$$\sigma_{\bar{x}} = \frac{1.8}{\sqrt{40}} = 0.285$$

Properties of the Normal Curve

Known characteristics of the normal curve make it possible to estimate the probability of occurrence of any value of a normally distributed variable. Suppose that the total area under the curve is defined to be 1. You can multiply that number by 100 and say there is a 100 percent chance that any value you can name will be somewhere in the distribution. (***Remember:*** The distribution extends to infinity in both directions.) Similarly, because half the area of the curve is below the mean and half is above it, you can say that there is a 50 percent chance that a randomly chosen value will be above the mean and the same chance that it will be below it.

It makes sense that the area under the normal curve is equivalent to the probability of randomly drawing a value in that range. The area is greatest in the middle, where the "hump" is, and thins out toward the tails. That is consistent with the fact that there are more values close to the mean in a normal distribution than far from it.

When the area of the standard normal curve is divided into sections by standard deviations above and below the mean, the area in each section is a known quantity (see Figure 5-4). As explained earlier, the area in each section is the same as the probability of randomly drawing a value in that range.

Figure 5-4 The normal curve and the area under the curve between σ units.

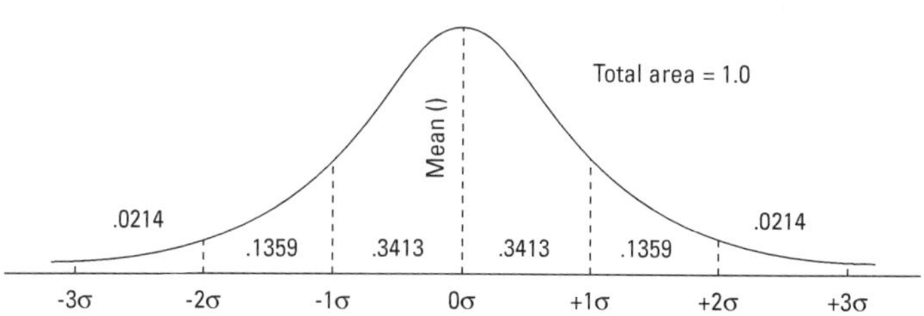

For example, 0.3413 of the curve falls between the mean and one standard deviation above the mean, which means that about 34 percent of all the values of a normally distributed variable are between the mean and one standard deviation above it. It also means that there is a 0.3413 chance that a value drawn at random from the distribution will lie between these two points.

Sections of the curve above and below the mean may be added together to find the probability of obtaining a value within (plus or minus) a given number of standard deviations of the mean (see Figure 5-5). For example, the amount of curve area between one standard deviation above the mean and one standard deviation below is 0.3413 + 0.3413 = 0.6826, which means that approximately 68.26 percent of the values lie in that range. Similarly, about 95 percent of the values lie within two standard deviations of the mean, and 99.7 percent of the values lie within three standard deviations.

Figure 5-5 The normal curve and the area under the curve between σ units.

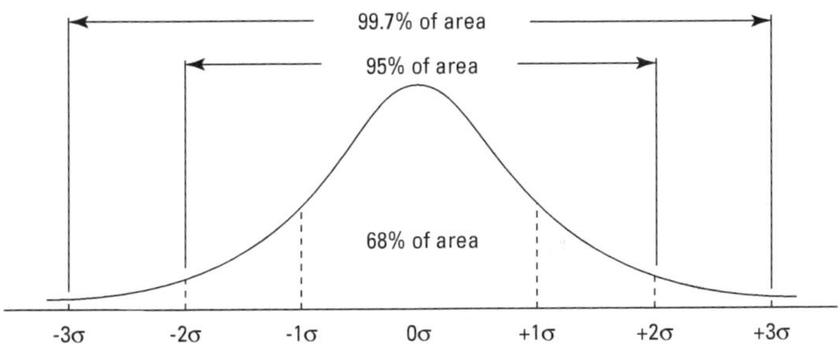

In order to use the area of the normal curve to determine the probability of occurrence of a given value, the value must first be **standardized,** or converted to a ***z*-score.** To convert a value to a *z*-score is to express it in terms of how many standard deviations it is above or below the mean. After the *z*-score is obtained, you can look up its corresponding probability in a table. The formula to compute a *z*-score is

$$z = \frac{x - \mu}{\sigma}$$

where x is the value to be converted, μ is the population mean, and σ is the population standard deviation.

Example 2: A normal distribution of retail-store purchases has a mean of \$14.31 and a standard deviation of 6.40. What percentage of purchases were under \$10? First, compute the *z*-score:

$$z = \frac{10 - 14.31}{6.40} = -0.67$$

The next step is to look up the *z*-score in the table of standard normal probabilities (see Table 2 in Appendix B). The standard normal table lists the probabilities (curve areas) associated with given *z*-scores.

Table 2 from Appendix B gives the area of the curve below *z*—in other words, the probability of obtaining a value of *z* or lower. Not all standard normal tables use the same format, however. Some list only positive *z*-scores and give the area of the curve between the mean and *z*. Such a table is slightly more difficult to use, but the fact that the normal curve is symmetric makes it possible to use it to determine the probability associated with any *z*-score, and vice versa.

To use Table 2 (the table of standard normal probabilities) in Appendix B, first look up the *z*-score in the left column, which lists *z* to the first decimal place. Then look along the top row for the second decimal place. The intersection of the row and column is the probability. In the example, you first find –0.6 in the left column and then 0.07 in the top row. Their intersection is 0.2514. The answer, then, is that about 25 percent of the purchases were under \$10 (see Figure 5-6).

Figure 5-6 Finding a probability using a *z*-score on the normal curve.

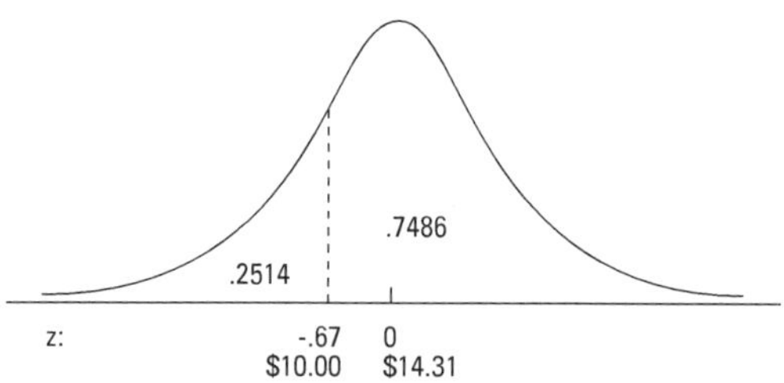

What if you had wanted to know the percentage of purchases above a certain amount? Because Table 2 gives the area of the curve below a given *z*, to obtain the area of the curve above *z*, simply subtract the tabled probability from 1. The area of the curve above a *z* of –0.67 is 1 – 0.2514 = 0.7486. Approximately 75 percent of the purchases were above \$10.

Just as Table 2 can be used to obtain probabilities from *z*-scores, it can be used to do the reverse.

Example 3: Using the previous example, what purchase amount marks the lower 10 percent of the distribution?

Locate in Table 2 the probability of 0.1000, or as close as you can find, and read off the corresponding *z*-score. The figure that you seek lies between the tabled probabilities of 0.0985 and 0.1003, but closer to 0.1003, which corresponds to a *z*-score of –1.28. Now, use the *z* formula, this time solving for *x*:

$$\frac{x-14.31}{6.4} = -1.28$$

$$x - 14.31 = (-1.28)(6.4)$$

$$x = -8.192 + 14.31$$

$$= 6.118$$

Approximately 10 percent of the purchases were below \$6.12.

Normal Approximation to the Binomial

In Chapter 4, which discusses probability, you saw that some variables are continuous—there is no limit to the number of times you could divide their intervals into still smaller ones, although you may round them off for convenience. Examples include age, height, and cholesterol level. Other variables are discrete, or made of whole units with no values between them. Some discrete variables are the number of children in a family, the sizes of televisions available for purchase, or the number of medals awarded at the Olympic Games.

Chapter 4 also introduces one kind of discrete variable: the binomial variable. A binomial variable can take only two values, often termed *successes* and *failures.* Examples include coin tosses that come up either heads or tails, manufactured parts that either continue working past a certain point or do not, and basketball tosses that either fall through the hoop or do not.

You discovered that the outcomes of binomial trials have a frequency distribution, just as continuous variables do. The more binomial trials there are (for example, the more coins you toss simultaneously), the more closely the sampling distribution resembles a normal curve (see Figure 5-7). You can take advantage of this fact and use the table of standard normal probabilities (Table 2 in Appendix B) to estimate the likelihood of obtaining a given proportion of successes. You can do this by converting the test proportion to a z-score and looking up its probability in the standard normal table.

Figure 5-7 As the number of trials increases, the binomial distribution approaches the normal distribution.

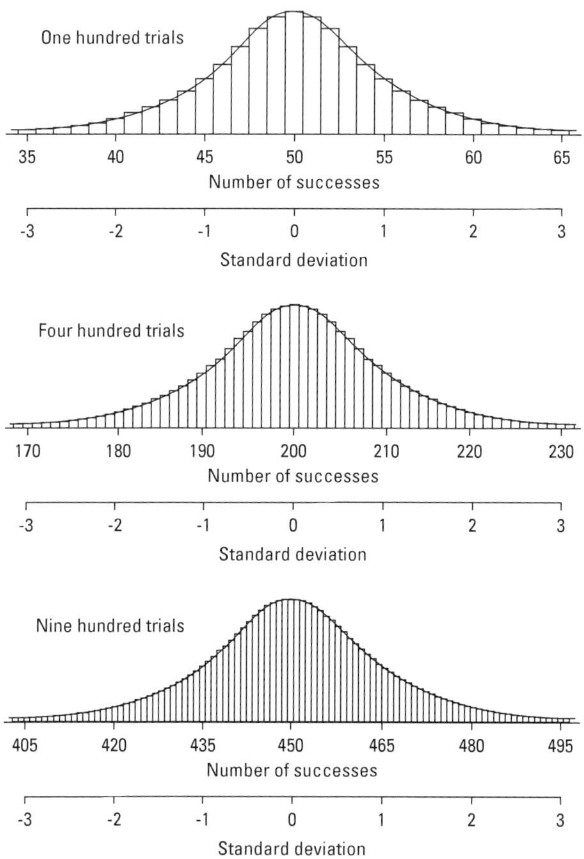

The mean of the normal approximation to the binomial is

$$\mu = n\pi$$

and the standard deviation is

$$\sigma = \sqrt{n\pi(1-\pi)}$$

where n is the number of trials and π is the probability of success. The approximation will be more accurate the larger the n and the closer the proportion of successes in the population to 0.5.

Example 4: Assuming an equal chance of a new baby being a boy or a girl (that is, $\pi = 0.5$), what is the likelihood that more than 60 out of the next 100 births at a local hospital will be boys?

$$\begin{aligned} z &= \frac{x-\mu}{\sigma} \\ &= \frac{60-((100)(0.5))}{\sqrt{(100)(0.5)(1-0.5)}} \\ &= \frac{10}{5} \\ &= 2 \end{aligned}$$

According to Table 2, a z-score of 2 corresponds to a probability of 0.9772. As you can see in Figure 5-8, there is a 0.9772 chance that there will be 60 percent or fewer boys, which means that the probability that there will be more than 60 percent boys is 1 – 0.9772 = 0.0228, or just over 2 percent. If the assumption that the chance of a new baby being a girl is the same as it being a boy is correct, the probability of obtaining 60 or fewer girls in the next 100 births is also 0.9772.

Figure 5-8 Finding a probability using a *z*-score on the normal curve.

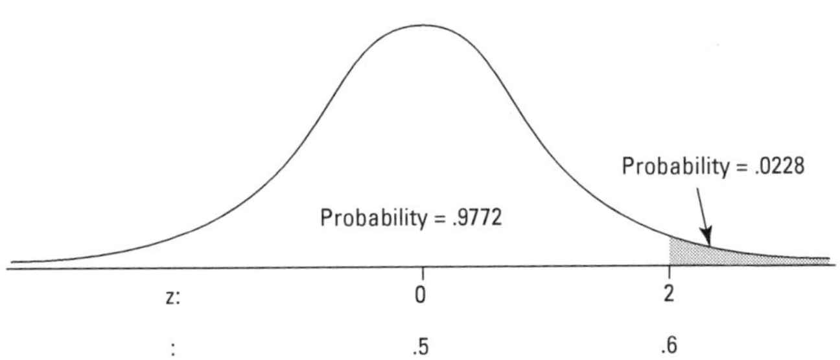

Chapter Check-Out

Q&A

1. A population has a mean of 500 and a standard deviation of 100. What is the probability that a randomly selected sample from this population, of size 25, will have a mean greater than 530?

2. The speed limit on the street in front of your house is 40 mph. In order to measure compliance with the limit, you decide to measure the mean speed of vehicles on the street. Each day after work, you measure the speed of 20 cars and take their average. Your data for five days is shown in the following table.

Day	*Average Speed (mph)*
Monday	38.3
Tuesday	40.1
Wednesday	34.2
Thursday	45.1
Friday	50.7

Using this information answer the following questions:

a. Find the mean and standard deviation of these means. Is it likely that drivers comply, on average, with the law?

b. Assuming the population mean of drivers does comply, what is the probability of your results? Take $\sigma = 7.5$ mph. Assume a normal distribution of speeds about μ.

Answers: 1. 0.0668 **2a.** 41.68, 6.38, yes. **b.** 0.59

Critical Thinking

3. On TV, surveys are often taken during sensational news stories to determine the "people's opinion." Usually, viewers are told to dial a phone number in order to vote yes or no on the issue. These phone numbers sometimes require the caller to pay a fee. Comment on this practice, in terms of sampling practice. Does it produce a random sample? If not, how could it be modified to produce a more random sample? Who is most likely to call these numbers?

Chapter 6
PRINCIPLES OF TESTING

Chapter Check-In

- ❑ Stating hypotheses and corresponding null hypotheses
- ❑ Learning how to apply simple statistical tests to determine if a null hypothesis is rejected
- ❑ Understanding the types of statistical errors and estimating them
- ❑ Calculating test statistics, confidence intervals, and significance

Most people assume that the whole point of statistical analysis is to prove or disprove a certain claim. It is not usually possible to prove or disprove the claim, but it is possible to establish evidence to test a claim. This chapter introduces you to techniques for testing claims based on the available data. These techniques are simple yet surprisingly powerful.

Stating Hypotheses

One common use of statistics is the testing of scientific hypotheses. First, the investigator forms a **research hypothesis** that states an expectation to be tested. Then the investigator derives a statement that is the opposite of the research hypothesis. This statement is called the **null hypothesis** (in notation: H_0). It is the null hypothesis that is actually tested, not the research hypothesis. If the null hypothesis can be rejected, that is taken as evidence in favor of the research hypothesis (also called the **alternative hypothesis,** H_a in notation). Because individual tests are rarely conclusive, it is usually not said that the research hypothesis has been "proved," only that it has been supported.

An example of a research hypothesis comparing two groups might be the following:

> Fourth-graders in Elmwood School perform differently in math than fourth-graders in Lancaster School. This could be measured by comparing the means of these groups.
>
> Or in notation: $H_a: \mu_1 \neq \mu_2$
>
> or sometimes: $H_a: \mu_1 - \mu_2 \neq 0$

The null hypothesis would be:

> Fourth-graders in Elmwood School perform the same in math as fourth-graders in Lancaster School.
>
> In notation: $H_0: \mu_1 = \mu_2$
>
> or: $H_0: \mu_1 - \mu_2 = 0$

Some research hypotheses are more specific than that, predicting not only a difference but a difference in a particular direction. These are often described as one-sided tests:

> Fourth-graders in Elmwood School are *better* in math than fourth-graders in Lancaster School.
>
> In notation: $H_a: \mu_1 > \mu_2$
>
> or: $H_a: \mu_1 - \mu_2 > 0$

The Test Statistic

Hypothesis testing involves the use of distributions of known area, like the normal distribution, to estimate the probability of obtaining a certain value as a result of chance. The researcher is usually testing to see if the probability will be low because that means it is likely that the test result was not a mere coincidence but occurred because the researcher's theory is correct. It could mean, for example, that it is probably not just bad luck but faulty packaging equipment that caused you to get a box of raisin cereal with only five raisins in it.

Only two outcomes of a hypothesis test are possible: Either the null hypothesis is rejected, or it is not. You have seen that values from normally distributed populations can be converted to z-scores and their probabilities looked up in Table 2 in Appendix B (see Chapter 5 for a complete discussion). The z-score is one kind of **test statistic** that is used to determine the probability of obtaining a given value. In order to test hypotheses, you must decide in advance what number to use as a cutoff for whether the null hypothesis will be rejected. This number is sometimes called the **critical** or **tabled value** because it is looked up in a table. It represents the level of probability that you will use to test the hypothesis. If the computed test statistic has a smaller probability than that of the critical value, the null hypothesis will be rejected.

For example, suppose you want to test the theory that sunlight helps prevent depression. One hypothesis derived from this theory might be that hospital admission rates for depression in sunny regions of the country are lower than the national average. Suppose that you know the national annual admission rate for depression to be 17 per 10,000. You intend to take the mean of a sample of admission rates from hospitals in sunny parts of the country and compare it to the national average.

Your research hypothesis is:

> The mean annual admission rate for depression from the hospitals in sunny areas is less than 17 per 10,000.
>
> In notation: H_a: $\mu_1 < 17$ per 10,000

The null hypothesis is:

> The mean annual admission rate for depression from the hospitals in sunny areas is equal to 17 per 10,000.
>
> In notation: H_0: $\mu_1 = 17$ per 10,000

Your next step is to choose a probability level for the test. You know that the sample mean must be lower than 17 per 10,000 in order to reject the null hypothesis, but how much lower? You settle on a probability level of 5 percent. That is, if the mean admission rate for the sample of sunny hospitals is so low that the chance of obtaining that rate from a sample selected at random from the national population is less than 5 percent, you will reject the null hypothesis and conclude that there is evidence to support the hypothesis that exposure to the sun reduces the incidence of depression.

Next, you look up the critical *z*-score—the *z*-score that corresponds to your chosen level of probability—in the standard normal table. It is important to remember which end of the distribution you are concerned with. Table 2 in Appendix B lists the probability of obtaining a given *z*-score or lower. That is, it gives the area of the curve below the *z*-score. Because a computed test statistic in the lower end of the distribution will allow you to reject your null hypothesis, you look up the *z*-score for the probability (or area) of 0.05 and find that it is –1.65. If you were hypothesizing that the mean in sunny parts of the country is greater than the national average, you would have been concerned with the upper end of the distribution instead and would have looked up the *z*-score associated with the probability (area) of 0.95, which is $z = 1.65$.

The critical *z*-score allows you to define the **region of acceptance** and the **region of rejection** of the curve (see Figure 6-1). If the computed test statistic is below the critical *z*-score, you can reject the null hypothesis and say that you have provided evidence in support of the alternative hypothesis. If it is above the critical value, you cannot reject the null hypothesis.

Figure 6-1 The *z*-score defines the boundary of the zones of rejection and acceptance.

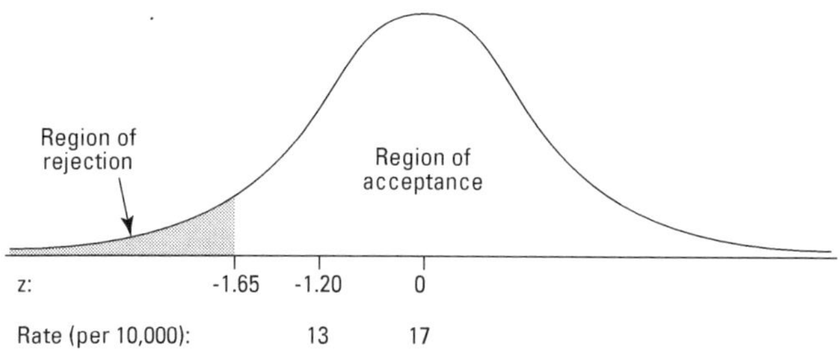

Suppose that the mean admission rate for the sample of hospitals in sunny regions is 13 per 10,000 and suppose also that the corresponding *z*-score for that mean is –1.20. The test statistic falls in the region of acceptance; so you cannot reject the null hypothesis that the mean in sunny parts of the country is significantly lower than the mean in the national average. There is a greater than 5 percent chance of obtaining a mean admission rate of 13 per 10,000 or lower from a sample of hospitals chosen at random from the national population, so you cannot conclude that your sample mean could not have come from that population.

One- and Two-Tailed Tests

In the previous example, you tested a research hypothesis that predicted not only that the sample mean would be different from the population mean but that it would be different in a specific direction—it would be lower. This test is called a **directional** or **one-tailed test** because the region of rejection is entirely within one tail of the distribution.

Some hypotheses predict only that one value will be different from another, without additionally predicting which will be higher. The test of such a hypothesis is **nondirectional** or **two-tailed** because an extreme test statistic in either tail of the distribution (positive or negative) will lead to the rejection of the null hypothesis of no difference.

Suppose that you suspect that a particular class's performance on a proficiency test is not representative of those people who have taken the test. The national mean score on the test is 74.

The research hypothesis is:

> The mean score of the class on the test is not 74.
>
> Or in notation: H_a: $\mu \neq 74$

The null hypothesis is:

> The mean score of the class on the test is 74.
>
> In notation: H_0: $\mu = 74$

As in the last example, you decide to use a 5 percent probability level for the test. Both tests have a region of rejection, then, of 5 percent, or 0.05. In this example, however, the rejection region must be split between both tails of the distribution—0.025 in the upper tail and 0.025 in the lower tail—because your hypothesis specifies only a difference, not a direction, as shown in Figure 6-2(a). You will reject the null hypotheses of no difference if the class sample mean is either much higher or much lower than the population mean of 74. In the previous example, only a sample mean much lower than the population mean would have led to the rejection of the null hypothesis.

Figure 6-2 Comparison of (a) a two-tailed test and (b) a one-tailed test, at the same probability level (95 percent).

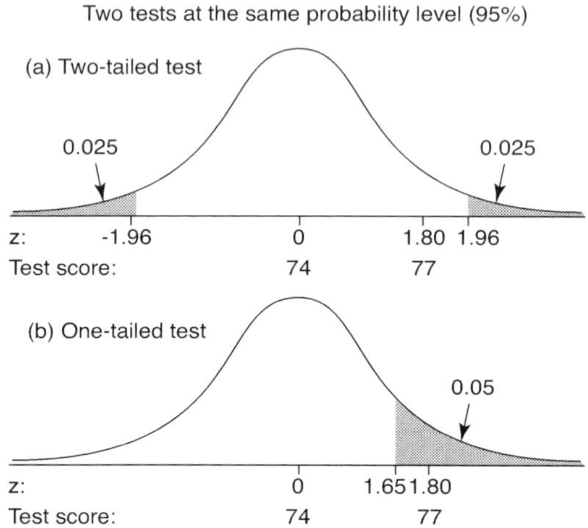

The decision of whether to use a one- or a two-tailed test is important because a test statistic that falls in the region of rejection in a one-tailed test may not do so in a two-tailed test, even though both tests use the same probability level. Suppose the class sample mean in your example was 77, and its corresponding z-score was computed to be 1.80. Table 2 in Appendix B shows the critical z-scores for a probability of 0.025 in either tail to be -1.96 and 1.96. In order to reject the null hypothesis, the test statistic must be either smaller than -1.96 or greater than 1.96. It is not, so you cannot reject the null hypothesis. Refer to Figure 6-2(a).

Suppose, however, you had a reason to expect that the class would perform better on the proficiency test than the population, and you did a one-tailed test instead. For this test, the rejection region of 0.05 would be entirely within the upper tail. The critical z-value for a probability of 0.05 in the upper tail is 1.65. (Remember that Table 2 in Appendix B gives areas of the curve below z; so you look up the z-value for a probability of 0.95.) Your computed test statistic of $z = 1.80$ exceeds the critical value and falls in the region of rejection, so you reject the null hypothesis and say that your suspicion that the class was better than the population was supported. See Figure 6-2(b).

In practice, you should use a one-tailed test only when you have good reason to expect that the difference will be in a particular direction. A two-tailed test is more conservative than a one-tailed test because a two-tailed test takes a more extreme test statistic to reject the null hypothesis.

Type I and II Errors

You have been using probability to decide whether a statistical test provides evidence for or against your predictions. If the likelihood of obtaining a given test statistic from the population is very small, you reject the null hypothesis and say that you have supported your hunch that the sample you are testing is different from the population.

But you could be wrong. Even if you choose a probability level of 5 percent, that means there is a 5 percent chance, or 1 in 20, that you rejected the null hypothesis when it was, in fact, correct. You can err in the opposite way, too; you might fail to reject the null hypothesis when it is, in fact, incorrect. These two errors are called Type I and Type II, respectively. Table 6-1 presents the four possible outcomes of any hypothesis test based on (1) whether the null hypothesis was accepted or rejected and (2) whether the null hypothesis was true in reality.

Table 6-1 Types of Statistical Errors

	H_0 is actually:	
	True	***False***
Reject H_0	Type I error	Correct
Accept H_0	Correct	Type II error

A **Type I error** is often represented by the Greek letter alpha (α) and a Type II error by the Greek letter beta (β). In choosing a level of probability for a test, you are actually deciding how much you want to risk committing a Type I error—rejecting the null hypothesis when it is, in fact, true. For this reason, the area in the region of rejection is sometimes called the alpha level because it represents the likelihood of committing a Type I error.

In order to graphically depict a Type II, or β, error, it is necessary to imagine next to the distribution for the null hypothesis a second distribution for the true alternative (see Figure 6-3). If the alternative hypothesis is actually true, but you fail to reject the null hypothesis for all values of the test statistic falling to the left of the critical value, then the area of the curve of the alternative (true) hypothesis lying to the left of the critical value represents the percentage of times that you will have made a Type II error.

Figure 6-3 Graphical depiction of the relation between Type I and Type II errors, and the power of the test.

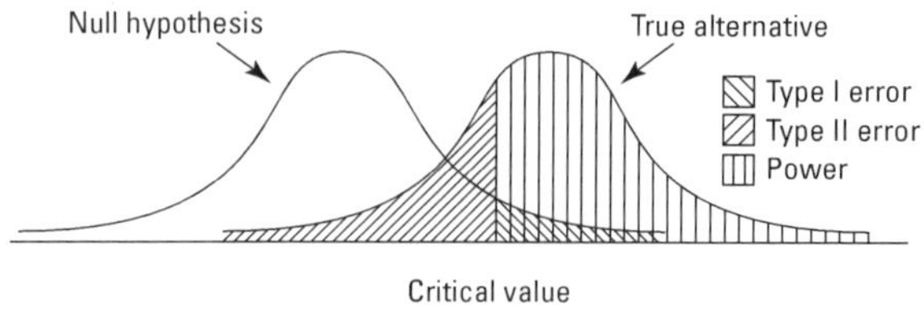

Type I and Type II errors are inversely related: As one increases, the other decreases. The Type I, or α (alpha), error rate is usually set in advance by the researcher. The Type II error rate for a given test is harder to know because it requires estimating the distribution of the alternative hypothesis, which is usually unknown.

A related concept is **power**—the probability that a test will reject the null hypothesis when it is, in fact, false. You can see from Figure 6-3 that power is simply 1 minus the Type II error rate (β). High power is desirable. Like β, power can be difficult to estimate accurately, but increasing the sample size always increases power.

Significance

How do you know how much confidence to put in the outcome of a hypothesis test? The statistician's criterion is the **statistical significance** of the test, or the likelihood of obtaining a given result by chance. This concept has been spoken of already, using several terms: probability, area of the curve, Type I error rate, and so forth. Another common representation of significance is the letter *p* (for probability) and a number between 0 and 1. There are several ways to refer to the significance level of a test, and it is important to be familiar with them. All of the following statements, for example, are equivalent:

The finding is significant at the 0.05 level.

The confidence level is 95 percent.

The Type I error rate is 0.05.

The alpha level is 0.05.

$\alpha = 0.05$.

There is a 1 in 20 chance of obtaining this result (or one more extreme).

The area of the region of rejection is 0.05.

The p-value is 0.05.

$p = 0.05$.

The smaller the significance level p, the more stringent the test and the greater the likelihood that the conclusion is correct. The significance level usually is chosen in consideration of other factors that affect and are affected by it, like sample size, estimated size of the effect being tested, and consequences of making a mistake. Common significance levels are 0.10 (1 chance in 10), 0.05 (1 chance in 20), and 0.01 (1 chance in 100).

The result of a hypothesis test, as has been seen, is that the null hypothesis is either rejected or not. The significance level for the test is set in advance by the researcher in choosing a critical test value. When the computed test statistic is large (or small) enough to reject the null hypothesis, however, it is customary to report the observed (actual) p-value for the statistic.

If, for example, you intend to perform a one-tailed (lower tail) test using the standard normal distribution at $p = 0.05$, the test statistic will have to be smaller than the critical z-value of -1.65 in order to reject the null hypothesis. But suppose the computed z-score is -2.50, which has an associated probability of 0.0062. The null hypothesis is rejected with room to spare. The observed significance level of the computed statistic is $p = 0.0062$; so you could report that the result was significant at $p < 0.01$. This result means that even if you had chosen the more stringent significance level of 0.01 in advance, you still would have rejected the null hypothesis, which is stronger support for your research hypothesis than rejecting the null hypothesis at $p = 0.05$.

It is important to realize that statistical significance and substantive, or practical, significance are not the same thing. A small, but important, real-world difference may fail to reach significance in a statistical test. Conversely, a statistically significant finding may have no practical consequence. This finding is especially important to remember when working with large sample sizes because any difference can be statistically significant if the samples are extremely large.

Point Estimates and Confidence Intervals

You have seen that the sample mean $\bar{x}$ is an unbiased estimate of the population mean μ. Another way to say this is that $\bar{x}$ is the best point estimate of the true value of μ. Some error is associated with this estimate, however—the true population mean may be larger or smaller than the sample mean. Instead of a point estimate, you might want to identify a range of possible values *p* might take, controlling the probability that μ is not lower than the lowest value in this range and not higher than the highest value. Such a range is called a **confidence interval.**

Example 1: Suppose that you want to find out the average weight of all players on the football team at Landers College. You are able to select ten players at random and weigh them. The mean weight of the sample of players is 198, so that number is your point estimate. Assume that the population standard deviation is $\sigma = 11.50$. What is a 90 percent confidence interval for the population weight, if you presume the players' weights are normally distributed?

This question is the same as asking what weight values correspond to the upper and lower limits of an area of 90 percent in the center of the distribution. You can define that area by looking up in Table 2 (in Appendix B) the *z*-scores that correspond to probabilities of 0.05 in either end of the distribution. They are –1.65 and 1.65. You can determine the weights that correspond to these *z*-scores using the following formula:

$$(a,b) = \bar{x} \pm z \cdot \frac{\sigma}{\sqrt{n}}$$

The weight values for the lower and upper ends of the confidence interval are 192 and 204 (see Figure 6-4). A confidence interval is usually expressed by two values enclosed by parentheses, as in (192, 204). Another way to express the confidence interval is as the point estimate plus or minus a margin of error; in this case, it is 198 ± 6 pounds. You are 90 percent certain that the true population mean of football player weights is between 192 and 204 pounds.

Figure 6-4 The relationship between point estimate, confidence interval, and *z*-score.

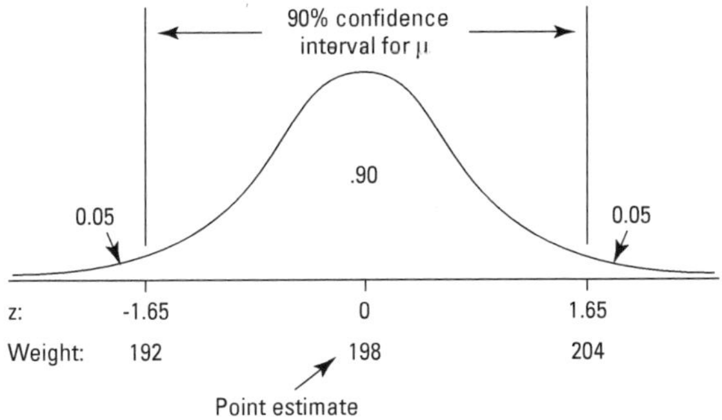

What would happen to the confidence interval if you wanted to be 95 percent certain of it? You would have to draw the limits (ends) of the intervals closer to the tails, in order to encompass an area of 0.95 between them instead of 0.90. That would make the low value lower and the high value higher, which would make the interval wider. The width of the confidence interval is related to the confidence level, standard error, and *n* such that the following are true:

- The higher the percentage of confidence desired, the wider the confidence interval.
- The larger the standard error, the wider the confidence interval.
- The larger the *n*, the smaller the standard error, and so the narrower the confidence interval.

All other things being equal, a smaller confidence interval is always more desirable than a larger one because a smaller interval means the population parameter can be estimated more accurately.

Estimating a Difference Score

Imagine that instead of estimating a single population mean μ, you wanted to estimate the difference between two population means μ_1 and μ_2, such as the difference between the mean weights of two football teams. The statistic $\bar{x}_1 - \bar{x}_2$ has a sampling distribution just as the individual means do, and the rules of statistical inference can be used to calculate either a point estimate or a confidence interval for the difference between the two population means.

Suppose you wanted to know which was greater, the mean weight of Landers College's football team or the mean weight of Ingram College's team. You already have a point estimate of 198 pounds for Landers's team. Suppose that you draw a random sample of players from Ingram's team, and the sample mean is 195. The point estimate for the difference between the mean weights of Landers's team (μ_1) and Ingram's team (μ_2) is 198 – 195 = 3.

But how accurate is that estimate? You can use the sampling distribution of the difference score to construct a confidence interval for $\mu_1 - \mu_2$, just as you did for Figure 6-4. Suppose that when you do so, you find that the confidence interval limits are (–3, 9), which means that you are 90 percent certain that the mean for the Landers team is between 3 pounds lighter and 9 pounds heavier than the mean for the Ingram team (see Figure 6-5).

Figure 6-5 The relationship between point estimate, confidence interval, and *z*-score, for a test of the difference of two means.

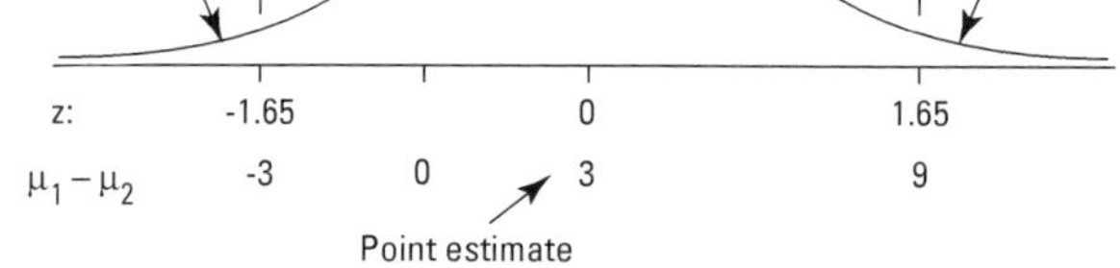

Suppose that instead of a confidence interval, you want to test the two-tailed hypothesis that the two team weights have different means. Your null hypothesis would be:

$$H_0: \mu_1 = \mu_2$$

or

$$H_0: \mu_1 - \mu_2 = 0$$

To reject the null hypothesis of equal means, the test statistic—in this example, *z*-score—for a difference in mean weights of 0 would have to fall in the rejection region at either end of the distribution. But you have already seen that it does not—only difference scores less than –3 or greater than 9 fall in the rejection region. For this reason, you would be unable to reject the null hypothesis that the two population means are equal.

This characteristic is a simple but important one of confidence intervals for difference scores. If the interval contains 0, you would be unable to reject the null hypothesis that the means are equal at the same significance level.

Univariate Tests: An Overview

To summarize, hypothesis testing of problems with one variable requires carrying out the following steps:

1. **State the null hypothesis and the alternative hypothesis.**
2. **Decide on a significance level for the test.**
3. **Compute the value of a test statistic.**
4. **Compare the test statistic to a critical value from the appropriate probability distribution corresponding to your chosen level of significance and observe whether the test statistic falls within the region of acceptance or the region of rejection. Equivalently, compute the *p*-value that corresponds to the test statistic and compare it to the selected significance level.**

Thus far, you have used the test statistic z and the table of standard normal probabilities (Table 2 in Appendix B) to carry out your tests. There are other test statistics and other probability distributions. The general formula for computing a test statistic for making an inference about a single population is

$$\text{test statistic} = \frac{\text{observed sample statistic} - \text{hypothesized value}}{\text{standard error}}$$

where *observed sample statistic* is the statistic of interest from the sample (usually the mean), *hypothesized value* is the hypothesized population parameter (again, usually the mean), and *standard error* is the standard deviation of the sampling distribution divided by the positive square root of n.

The general formula for computing a test statistic for making an inference about a difference between two populations is

$$\text{test statistic} = \frac{\text{statistic}_1 - \text{stastistic}_2 - \text{hypothesized value}}{\text{standard error}}$$

where *statistic*$_1$ and *statistic*$_2$ are the statistics from the two samples (usually the means) to be compared, *hypothesized value* is the hypothesized difference between the two population parameters (0 if testing for equal values), and *standard error* is the standard error of the sampling distribution, whose formula varies according to the type of problem.

The general formula for computing a confidence interval is

$$\text{observed sample statistic} \pm \text{critical value} \times \text{standard error}$$

where *observed sample statistic* is the point estimate (usually the sample mean), *critical value* is from the table of the appropriate probability distribution (upper or positive value if z) corresponding to half the desired alpha level, and *standard error* is the standard error of the sampling distribution.

Why must the alpha level be halved before looking up the critical value when computing a confidence interval? Because the rejection region is split between both tails of the distribution, as in a two-tailed test. For a confidence interval at $\alpha = 0.05$, you would look up the critical value corresponding to an upper-tailed probability of 0.025.

Chapter Check-Out

Q&A

1. An advertiser wants to know if the average age of people watching a particular TV show regularly is less than 24 years.

   a. State the alternative and null hypotheses.

   b. Is this a one- or two-tailed test?

   c. A random survey of 50 viewers determines that their mean age is 19 years, with a standard deviation of 1.7 years. Find the 90 percent confidence interval of the age of the viewers.

   d. What is the significance level for rejecting the null hypothesis?

2. True or False: Statistical tests should always be performed on a null hypothesis.

3. True or False: A result with a high level of significance is always very important.

Answers: 1a. H_a: $\mu < 24$; H_0: $\mu \geq 24$ **b.** one **c.** (18.6, 19.4) **d.** 0.0016 **2.** True **3.** False

Critical Thinking

1. Discuss the relationship between statistical tests and the assumption of causality. For example, suppose a researcher shows that, to a level $\alpha = 0.001$, the students at Middletown High are shorter than the students at Clearview High. Would you say that going to Middletown High causes you to be short? What, if anything, would you conclude if research showed, to $\alpha = 0.001$, that people living near a power plant die at a younger age than people living elsewhere?

Chapter 7
UNIVARIATE INFERENTIAL TESTS

Chapter Check-In

- ❑ Learning about several statistical tests, including the *z*-test and *t*-test
- ❑ Calculating confidence levels and confidence intervals for each test
- ❑ Applying these tests to examples

Chapter 6 explained how to formulate hypotheses and test them with the *z*-test. In Chapter 6, you also saw how a general test statistic is constructed and can be used to test a hypothesis against any probability distribution.

In this chapter, you take a closer look at some of the most common statistical tests used in scientific and sociological research: the *z*- and *t*-tests. These tests are used to test hypotheses concerning population means or proportions. Each test is explained in a separate section with a simple format so that you can easily look up how to calculate the appropriate test statistic. The formula for finding the confidence interval of your data is given as well. Most useful of all, each section contains several simple examples to illustrate how to apply the test to your data.

One-Sample *z*-test

Requirements: Normally distributed population, σ known

Test for population mean

Hypothesis test

Formula: $z = \dfrac{\bar{x} - \Delta}{\dfrac{\sigma}{\sqrt{n}}}$

where $\bar{x}$ is the sample mean, Δ is a specified value to be tested, σ is the population standard deviation, and n is the size of the sample. Look up the significance level of the z-value in the standard normal table (Table 2 in Appendix B).

Example 1 (one-tailed test): A herd of 1,500 steer was fed a special high-protein grain for a month. A random sample of 29 were weighed and had gained an average of 6.7 pounds. If the standard deviation of weight gain for the entire herd is 7.1, test the hypothesis that the average weight gain per steer for the month was more than 5 pounds.

null hypothesis: H_0: $\mu = 5$

alternative hypothesis: H_a: $\mu > 5$

$$z = \frac{6.7 - 5}{\frac{7.1}{\sqrt{29}}} = \frac{1.7}{1.318} = 1.289$$

Tabled value for $z \leq 1.28$ is 0.8997

$$1 - 0.8997 = 0.1003$$

So, the conditional probability that a sample from the herd gains at least 6.7 pounds per steer is $p = 0.1003$. Should the null hypothesis of a weight gain of less than 5 pounds for the population be rejected? That depends on how conservative you want to be. If you had decided beforehand on a significance level of $p < 0.05$, the null hypothesis could not be rejected.

Example 2 (two-tailed test): In national use, a vocabulary test is known to have a mean score of 68 and a standard deviation of 13. A class of 19 students takes the test and has a mean score of 65.

Is the class typical of others who have taken the test? Assume a significance level of $p < 0.05$.

There are two possible ways that the class may differ from the population. Its scores may be lower than, or higher than, the population of all students taking the test; therefore, this problem requires a two-tailed test. First, state the null and alternative hypotheses:

null hypothesis: H_0: $\mu = 68$

alternative hypothesis: H_a: $\mu \neq 68$

Because you have specified a significance level, you can look up the critical z-value in Table 2 of Appendix B before computing the statistic. This is a two-tailed test; so the 0.05 must be split such that 0.025 is in the upper tail and another 0.025 in the lower. The z-value that corresponds to –0.025 is –1.96, which is the lower critical z-value. The upper value corresponds to 1 – 0.025, or 0.975, which gives a z-value of 1.96. The null hypothesis of no difference will be rejected if the computed z statistic falls outside the range of –1.96 to 1.96.

Next, compute the z statistic:

$$z = \frac{65-68}{\frac{13}{\sqrt{19}}} = \frac{-3}{2.982} = -1.006$$

Because –1.006 is between –1.96 and 1.96, the null hypothesis of population mean is 68 and cannot be rejected. That is, there is not evidence that this class can be considered different from others who have taken the test.

Confidence interval for population mean using *z*

Formula: $(a,\ b) = \bar{x} \pm z_{\alpha/2} \cdot \frac{\sigma}{\sqrt{n}}$

where a and b are the limits of the confidence interval, $\bar{x}$ is the sample mean, $z_{\alpha/2}$ is the upper (or positive) z-value from the standard normal table corresponding to half of the desired alpha level (because all confidence intervals are two-tailed), σ is the population standard deviation, and n is the size of the sample.

Example 3: A sample of 12 machine pins has a mean diameter of 1.15 inches, and the population standard deviation is known to be 0.04. What is a 99 percent confidence interval of diameter width for the population?

First, determine the z-value. A 99 percent confidence level is equivalent to $p < 0.01$. Half of 0.01 is 0.005. The z-value corresponding to an area of 0.005 is 2.58. The interval may now be calculated:

$$1.15 \pm 2.58 \cdot \frac{0.04}{\sqrt{12}} = 1.15 \pm 0.03$$

The interval is (1.12, 1.18).

We have 99 percent confidence that the population mean of pin diameters lies between 1.12 and 1.18 inches. Note that this is not the same as saying that 99 percent of the machine pins have diameters between 1.12 and 1.18 inches, which would be an incorrect conclusion from this test.

Choosing a sample size

Because surveys cost money to administer, researchers often want to calculate how many subjects will be needed to determine a population mean using a fixed confidence interval and significance level. The formula is

$$n = \left(\frac{2z_{\alpha/2}\sigma}{w} \right)^2$$

where n is the number of subjects needed, $z_{\alpha/2}$ is the critical z-value corresponding to the desired significance level, σ is the population standard deviation, and w is the desired confidence interval width.

Example 4: How many subjects will be needed to find the average age of students at Fisher College plus or minus a year, with a 95 percent significance level and a population standard deviation of 3.5?

$$n = \left(\frac{(2)(1.96)(3.5)}{2} \right)^2 = \left(\frac{13.72}{2} \right)^2 = 47.06$$

Rounding up, a sample of 48 students would be sufficient to determine students' mean age plus or minus one year. Note that the confidence interval width is always double the "plus or minus" figure.

One-Sample *t*-test

Requirements: Normally distributed population, σ is unknown

Test for population mean

Hypothesis test

Formula: $t = \dfrac{\bar{x} - \Delta}{\frac{s}{\sqrt{n}}}$

where $\bar{x}$ is the sample mean, Δ is a specified value to be tested, s is the sample standard deviation, and n is the size of the sample.

When the standard deviation of the sample is substituted for the standard deviation of the population, the statistic does not have a normal distribution; it has what is called the t-distribution (see Table 3 in Appendix B). Because there is a different t-distribution for each sample size, it is not practical to list a separate area-of-the-curve table for each one. Instead, critical t-values for common alpha levels (0.10, 0.05, 0.01, and so forth) are usually given in a single table for a range of sample sizes. For very large samples, the t-distribution approximates the standard normal (z) distribution. In practice, it is best to use t-distributions any time the population standard deviation is not known.

Values in the t-table are not actually listed by sample size but by degrees of freedom *(df)*. The number of degrees of freedom for a problem involving the t-distribution for sample size n is simply $n - 1$ for a one-sample mean problem.

Example 5 (one-tailed test): A professor wants to know if her introductory statistics class has a good grasp of basic math. Six students are chosen at random from the class and given a math proficiency test. The professor wants the class to be able to score above 70 on the test. The six students get scores of 62, 92, 75, 68, 83, and 95. Can the professor have 90 percent confidence that the mean score for the class on the test would be above 70?

null hypothesis: H_0: $\mu = 70$

alternative hypothesis: H_a: $\mu > 70$

First, compute the sample mean and standard deviation (see Chapter 2):

$$\begin{array}{r} 62 \\ 92 \\ 75 \\ 68 \\ 83 \\ +\ 95 \\ \hline 475 \end{array} \qquad \bar{x} = \frac{475}{6} = 79.17 \qquad s = 13.17$$

Next, compute the t-value:

$$t = \frac{79.17 - 70}{\frac{13.17}{\sqrt{6}}} = \frac{9.17}{5.38} = 1.71$$

To test the hypothesis, the computed t-value of 1.71 will be compared to the critical value in the t-table. But which do you expect to be larger and which do you expect to be smaller? One way to reason about this is to look at the formula and see what effect different means would have on the computation. If the sample mean had been 85 instead of 79.17, the resulting t-value would have been larger. Because the sample mean is in the numerator, the larger it is, the larger the resulting figure will be. At the same time, you know that a higher sample mean will make it more likely that the professor will conclude that the math proficiency of the class is satisfactory and that the null hypothesis of less-than-satisfactory class math knowledge can be rejected. Therefore, it must be true that the larger the computed t-value, the greater the chance that the null hypothesis can be rejected. It follows, then, that if the computed t-value is larger than the critical t-value from the table, the null hypothesis can be rejected.

A 90 percent confidence level is equivalent to an alpha level of 0.10. Because extreme values in one rather than two directions will lead to rejection of the null hypothesis, this is a one-tailed test, and you do not divide the alpha level by 2. The number of degrees of freedom for the problem is $6 - 1 = 5$. The value in the t-table for $t_{.10,5}$ is 1.476. Because the computed t-value of 1.71 is larger than the critical value in the table, the null hypothesis can be rejected, and the professor has evidence that the class mean on the math test would be at least 70.

Note that the formula for the one-sample t-test for a population mean is the same as the z-test, except that the t-test substitutes the sample standard deviation s for the population standard deviation σ and takes critical values from the t-distribution instead of the z-distribution. The t-distribution is particularly useful for tests with small samples ($n < 30$).

Example 6 (two-tailed test): A Little League baseball coach wants to know if his team is representative of other teams in scoring runs. Nationally, the average number of runs scored by a Little League team in a game is 5.7. He chooses five games at random in which his team scored 5, 9, 4, 11, and 8 runs. Is it likely that his team's scores could have come from the national distribution? Assume an alpha level of 0.05.

Because the team's scoring rate could be either higher than or lower than the national average, the problem calls for a two-tailed test. First, state the null and alternative hypotheses:

null hypothesis: H_0: $\mu = 5.7$

alternative hypothesis: H_a: $\mu \neq 5.7$

Next compute the sample mean and standard deviation:

$$\begin{array}{r} 5 \\ 9 \\ 4 \\ 11 \\ +\ 8 \\ \hline 37 \end{array} \qquad \bar{x} = \frac{37}{5} = 7.4 \qquad s = 2.88$$

Next, the t-value:

$$t = \frac{7.4 - 5.7}{\frac{2.88}{\sqrt{5}}} = \frac{1.7}{1.29} = 1.32$$

Now, look up the critical value from the t-table (Table 3 in Appendix B). You need to know two things in order to do this: the degrees of freedom and the desired alpha level. The degrees of freedom is 5 – 1 = 4. The overall alpha level is 0.05, but because this is a two-tailed test, the alpha level must be divided by two, which yields 0.025. The tabled value for $t_{.025,4}$ is 2.776. The computed t of 1.32 is smaller than the t from Table 3, so you cannot reject the null hypothesis that the mean of this team is equal to the population mean. The coach cannot conclude that his team is different from the national distribution on runs scored.

Confidence interval for population mean using *t*

Formula: $(a,b) = \bar{x} \pm t_{\alpha/2,df} \cdot \frac{s}{\sqrt{n}}$

where a and b are the limits of the confidence interval, $\bar{x}$ is the sample mean, $t_{\alpha/2,df}$ is the value from the t-table corresponding to half of the desired alpha level at $n - 1$ degrees of freedom, s is the sample standard deviation, and n is the size of the sample.

Example 7: Using the previous example, what is a 95 percent confidence interval for runs scored per team per game?

First, determine the t-value. A 95 percent confidence level is equivalent to an alpha level of 0.05. Half of 0.05 is 0.025. The t-value corresponding to an area of 0.025 at either end of the t-distribution for 4 degrees of freedom ($t_{.025,4}$) is 2.776. The interval may now be calculated:

$$(a,b) = 7.4 \pm 2.78\frac{2.88}{\sqrt{5}}$$
$$= 7.4 \pm 3.58$$
$$= (3.82, 10.98)$$

The interval is fairly wide, mostly because n is small.

Two-Sample *z*-test for Comparing Two Means

Requirements: Two normally distributed but independent populations, σ is known

Hypothesis test

Formula: $z = \dfrac{\bar{x}_1 - \bar{x}_2 - \Delta}{\sqrt{\dfrac{\sigma_1^2}{n_1} + \dfrac{\sigma_2^2}{n_2}}}$

where $\bar{x}_1$ and $\bar{x}_2$ are the means of the two samples, Δ is the hypothesized difference between the population means (0 if testing for equal means), σ_1 and σ_2 are the standard deviations of the two populations, and n_1 and n_2 are the sizes of the two samples.

Example 8 (two-tailed test): The amount of a certain trace element in blood is known to vary with a standard deviation of 14.1 ppm (parts per million) for male blood donors and 9.5 ppm for female donors. Random samples of 75 male and 50 female donors yield concentration means of 28 and 33 ppm, respectively. What is the likelihood that the population means of concentrations of the element are the same for men and women?

Null hypothesis: H_0: $\mu_1 = \mu_2$

or H_0: $\mu_1 - \mu_2 = 0$

alternative hypothesis: H_a: $\mu_1 \neq \mu_2$

or: H_a: $\mu_1 - \mu_2 \neq 0$

$$z = \frac{28 - 33 - 0}{\sqrt{\dfrac{14.1^2}{75} + \dfrac{9.5^2}{50}}} = \frac{-5}{\sqrt{2.65 + 1.81}} = -2.37$$

The computed z-value is negative because the (larger) mean for females was subtracted from the (smaller) mean for males. But because the hypothesized difference between the populations is 0, the order of the samples in this computation is arbitrary—$\overline{x}_1$ could just as well have been the female sample mean and $\overline{x}_2$ the male sample mean, in which case z would be 2.37 instead of –2.37. An extreme z-score in either tail of the distribution (plus or minus) will lead to rejection of the null hypothesis of no difference.

The area of the standard normal curve corresponding to a z-score of –2.37 is 0.0089. Because this test is two-tailed, that figure is doubled to yield a probability of 0.0178 that the population means are the same. If the test had been conducted at a pre-specified significance level of $\alpha < 0.05$, the null hypothesis of equal means could be rejected. If the specified significance level had been the more conservative (more stringent) $\alpha < 0.01$, however, the null hypothesis could not be rejected.

In practice, the two-sample z-test is not used often, because the two population standard deviations σ_1 and σ_2 are usually unknown. Instead, sample standard deviations and the t-distribution are used.

Two-Sample *t*-test for Comparing Two Means

Requirements: Two normally distributed but independent populations, σ is unknown

Hypothesis test

Formula: $t = \dfrac{\overline{x}_1 - \overline{x}_2 - \Delta}{\sqrt{\dfrac{s_1^2}{n_1} + \dfrac{s_2^2}{n_2}}}$

where $\overline{x}_1$ and $\overline{x}_2$ are the means of the two samples, Δ is the hypothesized difference between the population means (0 if testing for equal means), s_1 and s_2 are the standard deviations of the two samples, and n_1 and n_2 are the sizes of the two samples. The number of degrees of freedom for the problem is the smaller of $n_1 - 1$ and $n_2 - 1$.

Example 9 (one-tailed test): An experiment is conducted to determine whether intensive tutoring (covering a great deal of material in a fixed amount of time) is more effective than paced tutoring (covering less material in the same amount of time). Two randomly chosen groups are tutored separately and then administered proficiency tests. Use a significance level of $\alpha < 0.05$.

Let μ_1 represent the population mean for the intensive tutoring group and μ_2 represent the population mean for the paced tutoring group.

null hypothesis: H_0: $\mu_1 = \mu_2$

or H_0: $\mu_1 - \mu_2 = 0$

alternative hypothesis: H_a: $\mu_1 > \mu_2$

or: H_a: $\mu_1 - \mu_2 > 0$

Group	*Method*	*n*	$\bar{x}$	*s*
1	Intensive	12	46.31	6.44
2	Paced	10	42.79	7.52

$$t = \frac{46.31 - 42.79 - 0}{\sqrt{\frac{6.44^2}{12} + \frac{7.52^2}{10}}} = \frac{3.52}{\sqrt{3.46 + 5.66}} = 1.166$$

The degrees of freedom parameter is the smaller of (12 – 1) and (10 – 1), or 9. Because this is a one-tailed test, the alpha level (0.05) is not divided by two. The next step is to look up $t_{.05,9}$ in the *t*-table (Table 3 in Appendix B), which gives a critical value of 1.833. The computed *t* of 1.166 does not exceed the tabled value, so the null hypothesis cannot be rejected. This test has not provided statistically significant evidence that intensive tutoring is superior to paced tutoring.

Confidence interval for comparing two means

Formula: $(a,b) = \bar{x}_1 - \bar{x}_2 \pm t_{\alpha/2,df} \cdot \sqrt{\frac{s_1^2}{n_1} + \frac{s_2^2}{n_2}}$

where a and b are the limits of the confidence interval, $\bar{x}_1$ and $\bar{x}_2$ are the means of the two samples, $t_{\alpha/2,df}$ is the value from the *t*-table corresponding to half of the desired alpha level, s_1 and s_2 are the standard deviations of the two samples, and n_1 and n_2 are the sizes of the two samples. The degrees of freedom parameter for looking up the *t*-value is the smaller of $n_1 - 1$ and $n_2 - 1$.

Example 10: Estimate a 90 percent confidence interval for the difference between the number of raisins per box in two brands of breakfast cereal.

Brand	***n***	$\bar{x}$	***s***
A	6	102.1	12.3
B	9	93.6	7.52

The difference between $\bar{x}_1$ and $\bar{x}_2$ is 102.1 – 93.6 = 8.5. The degrees of freedom is the smaller of (6 – 1) and (9 – 1), or 5. A 90 percent confidence interval is equivalent to an alpha level of 0.10, which is then halved to give 0.05. According to Table 3 in Appendix B, the critical value for $t_{.05,5}$ is 2.015. The interval may now be computed.

$$8.5 \pm 2.015 \cdot \sqrt{\frac{12.3^2}{6} + \frac{7.52^2}{9}} = 8.5 \pm 2.015 \cdot \sqrt{25.22 + 6.28}$$
$$= 8.5 \pm 11.31$$

The interval is (–2.81, 19.81).

You can be 90 percent confident that Brand A cereal has between 2.81 fewer and 19.81 more raisins per box than Brand B. The fact that the interval contains 0 means that if you had performed a test of the hypothesis that the two population means are different (using the same significance level), you would not have been able to reject the null hypothesis of no difference.

Pooled variance method

If the two population distributions can be assumed to have the same variance—and, therefore, the same standard deviation—s_1 and s_2 can be pooled together, each weighted by the number of cases in each sample. Although using pooled variance in a *t*-test is generally more likely to yield significant results than using separate variances, it is often hard to know whether the variances of the two populations are equal. For this reason, the pooled variance method should be used with caution. The formula for the pooled estimator of σ^2 is

$$S_p^2 = \frac{(n_1 - 1)s_1^2 + (n_2 - 1)s_2^2}{n_1 + n_2 - 2}$$

where s_1 and s_2 are the standard deviations of the two samples and n_1 and n_2 are the sizes of the two samples.

The formula for comparing the means of two populations using pooled variance is

$$t = \frac{\bar{x}_1 - \bar{x}_2 - \Delta}{\sqrt{s_p^2\left(\frac{1}{n_1} + \frac{1}{n_2}\right)}}$$

where $\bar{x}_1$ and $\bar{x}_2$ are the means of the two samples, Δ is the hypothesized difference between the population means (0 if testing for equal means), s_p^2 is the pooled variance, and n_1 and n_2 are the sizes of the two samples. The number of degrees of freedom for the problem is

$$df = n_1 + n_2 - 2$$

Example 11 (two-tailed test): Does right- or left-handedness affect how fast people type? Random samples of students from a typing class are given a typing speed test (words per minute), and the results are compared. Significance level for the test: 0.10. Because you are looking for a difference between the groups in either direction (right-handed faster than left, or vice versa), this is a two-tailed test.

null hypothesis: H_0: $\mu_1 = \mu_2$

or: H_0: $\mu_1 - \mu_2 = 0$

alternative hypothesis: H_a: $\mu_1 \neq \mu_2$

or: H_a: $\mu_1 - \mu_2 \neq 0$

Group	***-Handed***	***n***	$\bar{x}$	***s***
1	Right	16	55.8	5.7
2	Left	9	59.3	4.3

First, calculate the pooled variance:

$$\begin{aligned} S_p^2 &= \frac{(16-1)5.7^2 + (9-1)4.3^2}{16+9-2} \\ &= \frac{487.35 + 147.92}{23} \\ &= 27.62 \end{aligned}$$

Next, calculate the *t*-value:

$$t = \frac{55.8 - 59.3 - 0}{\sqrt{27.62\left(\frac{1}{16} + \frac{1}{9}\right)}} = \frac{-3.5}{\sqrt{4.80}} = -1.598$$

The degrees-of-freedom parameter is 16 + 9 – 2, or 23. This test is a two-tailed one, so you divide the alpha level (0.10) by two. Next, you look up $t_{.05,23}$ in the *t*-table (Table 3 in Appendix B), which gives a critical value of 1.714. This value is larger than the absolute value of the computed *t* of –1.598, so the null hypothesis of equal population means cannot be rejected. There is no evidence that right- or left-handedness has any effect on typing speed.

Paired Difference *t*-test

Requirements: A set of paired observations from a normal population

This *t*-test compares one set of measurements with a second set from the same sample. It is often used to compare "before" and "after" scores in experiments to determine whether significant change has occurred.

Hypothesis test

Formula: $t = \dfrac{\bar{x} - \Delta}{\frac{s}{\sqrt{n}}}$

where $\bar{x}$ is the mean of the change scores, Δ is the hypothesized difference (0 if testing for equal means), s is the sample standard deviation of the differences, and n is the sample size. The number of degrees of freedom for the problem is $n - 1$.

Example 12 (one-tailed test): A farmer decides to try out a new fertilizer on a test plot containing 10 stalks of corn. Before applying the fertilizer, he measures the height of each stalk. Two weeks later, he measures the stalks again, being careful to match each stalk's new height to its previous one. The stalks would have grown an average of 6 inches during that time even without the fertilizer. Did the fertilizer help? Use a significance level of 0.05.

null hypothesis: H_0: $\mu = 6$

alternative hypothesis: H_a: $\mu > 6$

Stalk	*1*	*2*	*3*	*4*	*5*	*6*	*7*	*8*	*9*	*10*
Before height	35.5	31.7	31.2	36.3	22.8	28.0	24.6	26.1	34.5	27.7
After height	45.3	36.0	38.6	44.7	31.4	33.5	28.8	35.8	42.9	35.0

Subtract each stalk's "before" height from its "after" height to get the change score for each stalk; then compute the mean and standard deviation of the change scores and insert these into the formula.

$$\begin{aligned}
45.3-35.5&=9.8\\
36.0-31.7&=4.3\\
38.6-31.2&=7.4\\
44.7-36.3&=8.4\\
31.4-22.8&=8.6\\
33.5-28.0&=5.5\\
28.8-24.6&=4.2\\
35.8-26.1&=9.7\\
42.9-34.5&=8.4\\
35.0-27.7&=\underline{7.3}\\
\frac{73.6}{10}&=7.36
\end{aligned}$$

$$\begin{aligned}
s^2&=\frac{(9.8-7.36)^2+(4.3-7.36)^2+\ldots+(7.3-7.36)^2}{10-1}\\
&=4.216\\
s&=\sqrt{4.216}\\
&=2.05\\
t&=\frac{7.36-6}{\frac{2.05}{\sqrt{10}}}=\frac{1.36}{0.65}=2.09
\end{aligned}$$

The problem has $n - 1$, or $10 - 1 = 9$ degrees of freedom. The test is one-tailed because you are asking only whether the fertilizer increases growth, not reduces it. The critical value from the *t*-table for $t_{.05,9}$ is 1.833. Because the computed *t*-value of 2.098 is larger than 1.833, the null hypothesis can be rejected. The test has provided evidence that the fertilizer caused the corn to grow more than if it had not been fertilized. The amount of actual increase was not large (1.36 inches over normal growth), but it was statistically significant.

Test for a Single Population Proportion

Requirements: Binomial population, sample $n\pi_0 \geq 10$, and sample $n(1 - \pi_0) \geq 10$, where π_0 is the hypothesized proportion of successes in the population.

Hypothesis test

Formula: $z = \dfrac{\hat{\pi} - \pi_0}{\sqrt{\dfrac{\pi_0(1-\pi_0)}{n}}}$

where $\hat{\pi}$ is the sample proportion, π_0 is the hypothesized proportion, and n is the sample size. Because the distribution of sample proportions is approximately normal for large samples, the *z* statistic is used. The test is most accurate when π (the population proportion) is close to 0.5 and least accurate when π is close to 0 or 1.

Example 13 (one-tailed test): The sponsors of a city marathon have been trying to encourage more women to participate in the event. A sample of 70 runners is taken, of which 32 are women. The sponsors would like to be 90 percent certain that at least 40 percent of the participants are women. Were their recruitment efforts successful?

null hypothesis: H_0: $\pi = 0.4$

alternative hypothesis: H_0: $\pi > 0.4$

The proportion of women runners in the sample is 32 out of 70, or 45.7 percent. The *z*-value may now be calculated:

$$z = \frac{0.457 - 0.40}{\sqrt{\dfrac{0.40(1-0.40)}{70}}} = \frac{0.057}{\sqrt{0.003}} = 0.973$$

From the z-table, you find that the probability of a z-value less than 0.97 is 0.834, so we do not reject the null hypothesis, so it cannot be concluded at that level of significance that the population of runners is at least 40 percent women.

Confidence interval for a single population proportion

Formula: $(a,b)=\hat{\pi}\pm z_{\alpha/2}\cdot\sqrt{\frac{\hat{\pi}(1-\hat{\pi})}{n}}$

where $\hat{\pi}$ is the sample proportion, $z_{\alpha/2}$ is the upper z-value corresponding to half of the desired alpha level, and n is the sample size.

Example 14: A sample of 100 voters selected at random in a congressional district prefer Candidate Smith to Candidate Jones by a ratio of 3 to 2. What is a 95 percent confidence interval of the percentage of voters in the district who prefer Smith?

A ratio of 3 to 2 is equivalent to a proportion of $\frac{3}{5}=0.60$. A 95 percent confidence interval is equivalent to an alpha level of 0.05, half of which is 0.025. The critical z-value corresponding to an upper probability of 1 – 0.025 is 1.96. The interval may now be computed:

$$\begin{aligned}(a,b)&=0.60\pm1.96\sqrt{\frac{0.60(1-0.60)}{100}}\\&=0.60\pm1.96\sqrt{0.002}\\&=0.60\pm0.096\\&=(0.504,0.696)\end{aligned}$$

We have 95 percent confidence that between 50.4 percent and 69.6 percent of the voters in the district prefer Candidate Smith. Note that the problem could have been figured for Candidate Jones by substituting the proportion 0.40 for Smith's proportion of 0.60.

Choosing a sample size

In the previous problem, you estimated that the percentage of voters in the district who prefer Candidate Smith is 60 percent plus or minus about 10 percent. Another way to say this is that the estimate has a "margin of error" of ± 10 percent, or a confidence interval width of 20 percent. That is a pretty wide range. You may wish to make the margin smaller.

Because the width of the confidence interval decreases at a known rate as the sample size increases, it is possible to determine the sample size needed to estimate a proportion with a fixed confidence interval. The formula is

$$n=\left(\frac{2z_{\alpha/2}}{w}\right)^2\cdot p^*\left(1-p^*\right)$$

where n is the number of subjects needed, $z_{\alpha/2}$ is the z-value corresponding to half of the desired significance level, w is the desired confidence interval width, and p^* is an estimate of the true population proportion. A p^* of 0.50 will result in a higher n than any other proportion estimate but is often used when the true proportion is not known.

Example 15: How large a sample is needed to estimate the preference of district voters for Candidate Smith with a margin of error of $\pm$ 4 percent, at a 95 percent significance level?

You will conservatively estimate the (unknown) true population proportion of preference for Smith at 0.50. If it is really larger (or smaller) than that, you will overestimate the size of the sample needed, but $p^* = 0.50$ is playing it safe.

$$\begin{aligned} n &= \left(\frac{2\cdot 1.96}{0.08}\right)^2\cdot 0.50(1-0.50) \\ &= 49^2\cdot 0.25 \\ &= 600.25 \end{aligned}$$

A sample of about 601 voters would be needed to estimate the percentage of voters in the district who prefer Smith and to be 95 percent certain that the estimate is within $\pm$ 4 percent of the true population percentage.

Test for Comparing Two Proportions

Requirements: Two binomial populations, $n\,\pi_0 \geq 5$ and $n\,(1-\pi_0) \geq 5$ (for each sample), where π_0 is the hypothesized proportion of successes in the population.

Difference test

Hypothesis test

Formula: $z=\dfrac{\hat{\pi}_1-\hat{\pi}_2-\Delta}{\sqrt{\hat{\pi}\left(1-\hat{\pi}\right)\left(\dfrac{1}{n_1}+\dfrac{1}{n_2}\right)}}$

where $\hat{\pi} = \dfrac{x_1 + x_2}{n_1 + n_2}$

and where $\hat{\pi}_1$ and $\hat{\pi}_2$ are the sample proportions, Δ is their hypothesized difference (0 if testing for equal proportions), n_1 and n_2 are the sample sizes, and x_1 and x_2 are the number of "successes" in each sample. As in the test for a single proportion, the z distribution is used to test the hypothesis.

Example 16 (one-tailed test): A swimming school wants to determine whether a recently hired instructor is working out. Sixteen out of 25 of Instructor A's students passed the lifeguard certification test on the first try. In comparison, 57 out of 72 of more experienced Instructor B's students passed the test on the first try. Is Instructor A's success rate worse than Instructor B's? Use $\alpha = 0.10$.

null hypothesis: $H_0\colon \pi_1 = \pi_2$

alternative hypothesis: $H_a\colon \pi_1 < \pi_2$

First, you need to compute the values for some of the terms in the formula. The sample proportion $\hat{\pi}_1$ is $\dfrac{16}{25} = 0.640$. The sample proportion $\hat{\pi}_2$ is $\dfrac{57}{72} = 0.792$. Next, compute $\hat{\pi}$:

$$\hat{\pi} = \frac{16+57}{25+72} = \frac{73}{97} = 0.753$$

Finally, the main formula:

$$\begin{aligned} z &= \frac{0.640 - 0.792 - 0}{\sqrt{0.753(1-0.753)\left(\dfrac{1}{25} + \dfrac{1}{72}\right)}} \\ &= \frac{-0.152}{\sqrt{(0.186)(0.054)}} \\ &= \frac{-0.152}{0.100} \\ &= -1.518 \end{aligned}$$

The standard normal (z) table shows that the lower critical z-value for $\alpha =$ 0.10 is approximately -1.28. The computed z must be lower than -1.28 in order to reject the null hypothesis of equal proportions. Because the computed z is -1.518, the null hypothesis can be rejected. It can be concluded (at this level of significance) that Instructor A's success rate is worse than Instructor B's.

Confidence interval for comparing two proportions

Formula: $(a,b) = \hat{\pi}_1 - \hat{\pi}_2 \pm z_{\alpha/2} \cdot s(D)$

where $s(D) = \sqrt{\dfrac{\hat{\pi}_1(1-\hat{\pi}_1)}{n_1} + \dfrac{\hat{\pi}_2(1-\hat{\pi}_2)}{n_2}}$

and where a and b are the limits of the confidence interval of $\pi_1 - \pi_2$, $\hat{\pi}_1$ and $\hat{\pi}_2$ are the sample proportions, $z_{\alpha/2}$ is the upper z-value corresponding to half of the desired alpha level, and n_1 and n_2 are the sizes of the two samples.

Example 17: A public health researcher wants to know how two high schools—one in the inner city and one in the suburbs—differ in the percentage of students who smoke. A random survey of students gives the following results:

Population	*n*	*Smokers*
1 (inner-city)	125	47
2 (suburban)	153	52

What is a 90 percent confidence interval for the difference between the smoking rates in the two schools?

The proportion of smokers in the inner-city school is $\hat{\pi}_1 = \dfrac{47}{125} = 0.376.$

The proportion of smokers in the suburban school is $\hat{\pi}_2 = \dfrac{52}{153} = 0.340.$

Next solve for $s(D)$:

$$\begin{aligned} s(D) &= \sqrt{\frac{0.376(1-0.376)}{125} + \frac{0.340(1-0.340)}{153}} \\ &= \sqrt{\frac{0.235}{125} + \frac{0.224}{153}} \\ &= \sqrt{0.003} \\ &= 0.058 \end{aligned}$$

A 90 percent confidence interval is equivalent to $\alpha = 0.10$, which is halved to give 0.05. The upper tabled value for $z_{.05}$ is 1.65. The interval may now be computed:

$$\begin{aligned}(a,b) &= (0.376 - 0.340) \pm (1.65)(0.058)\\ &= 0.036 \pm 0.096\\ &= (-0.060, 0.132)\end{aligned}$$

The researcher can be 90 percent confident that the true population proportion of smokers in the inner-city high school is between 6 percent lower and 13.2 percent higher than the proportion of smokers in the suburban high school. Thus, since the confidence interval contains zero, there is no significant difference between the two types of schools at $\alpha = 0.10$.

Chapter Check-Out

Q&A

1. Suppose you count the number of days in May on which it rains in your town and find it to be 15 (out of 31). Checking the weather database, you find that for your town, the average number of rainy days in May is 10.
   - **a.** What type of test would you use to determine if this May was unusually wet to $\alpha = 0.05$?
   - **b.** Perform the test. What is the value of your test statistic (that is, z or t)?
2. A biologist believes that polluted water may be causing frogs in a lake to be smaller than normal. In an unpolluted lake, she measures 15 adult frogs and finds a mean length of 7.6 inches with a standard deviation of 1.5 inches. In the polluted lake, she measures 23 adult frogs and finds a mean of 6.5 inches and a standard deviation of 2.3 inches.
   - **a.** State the null and alternative hypotheses.
   - **b.** Calculate the appropriate test statistic.

3. A high school track coach makes the sprinters on his team begin weight lifting, in the hopes that their speed will increase. Before beginning, each runs a 40-yard dash. After a month of training, each runs the dash again. Their times are given below.

Runner	*1*	*2*	*3*	*4*	*5*	*6*	*7*	*8*	*9*	*10*
Before training	4.8	5.1	5.5	4.9	5.6	6.0	5.8	5.3	6.1	4.7
After training	4.7	5.0	5.1	5.0	5.2	5.1	5.6	5.1	5.5	5.0

a. State the alternative and null hypotheses.
b. To what level can the coach be sure that his runners have improved?

Answers: 1a. test for single population proportion **b.** 1.92 **2a.** H_a: $\mu_{pol} < \mu_{unpol}$; H_0: $\mu_{pol} \geq \mu_{unpol}$ **b.** 1.784 **3a.** H_a: $\mu_{after} < \mu_{before}$; H_0: $\mu_{after} \geq \mu_{before}$ **b.** $p < 0.025$

Chapter 8
BIVARIATE RELATIONSHIPS

Chapter Check-In

- ❑ Checking for correlation between two variables
- ❑ Learning about regression
- ❑ Using the chi-square (χ^2) test to determine whether two variables are independent

So far, you have been working with problems involving a single variable. Many times, however, you want to know something about the relationship between two variables.

Correlation

Consider Table 8-1, which contains measurements on two variables for ten people: the number of months the person has owned an exercise machine and the number of hours the person spent exercising in the past week.

Table 8-1 Exercise Data for Ten People

Person	*1*	*2*	*3*	*4*	*5*	*6*	*7*	*8*	*9*	*10*
Months Owned	5	10	4	8	2	7	9	6	1	12
Hours Exercised	5	2	8	3	8	5	5	7	10	3

If you display these data pairs as points in a scatter plot (see Figure 8-1), you can see a definite trend. The points appear to form a line that slants from the upper left to the lower right. As you move along that line from left to right, the values on the vertical axis (hours of exercise) get smaller, while the values on the horizontal axis (months owned) get larger.

Another way to express this is to say that the two variables are inversely related: The more months the machine was owned, the less the person tended to exercise.

Figure 8-1 The data in Table 8-1 is an example of negative correlation.

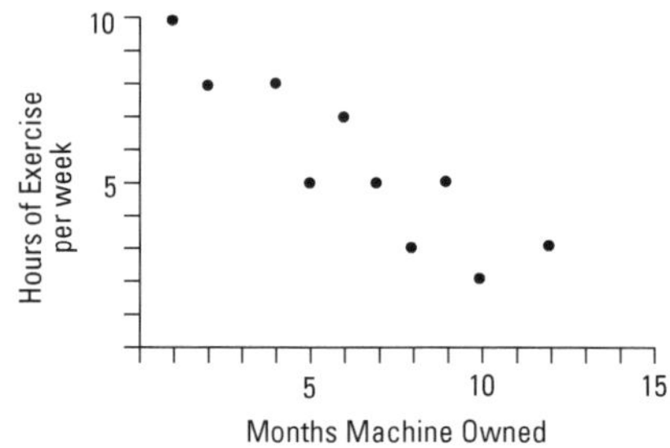

These two variables are **correlated.** More than that, they are correlated in a particular direction—negatively. For an example of a positive correlation, suppose that instead of displaying "hours of exercise" on the vertical axis, you put the person's score on a test that measures cardiovascular fitness (see Figure 8-2). The pattern of these data points suggests a line that slants from lower left to upper right, which is the opposite of the direction of slant in the first example. Figure 8-2 shows that the longer the person has owned the exercise machine, the better his or her cardiovascular fitness tends to be. This might be true in spite of the fact that time spent exercising decreases the longer the machine has been owned because purchasers of exercise machines might be starting from a point of low fitness, which may improve only gradually.

Figure 8-2 An example of positive correlation.

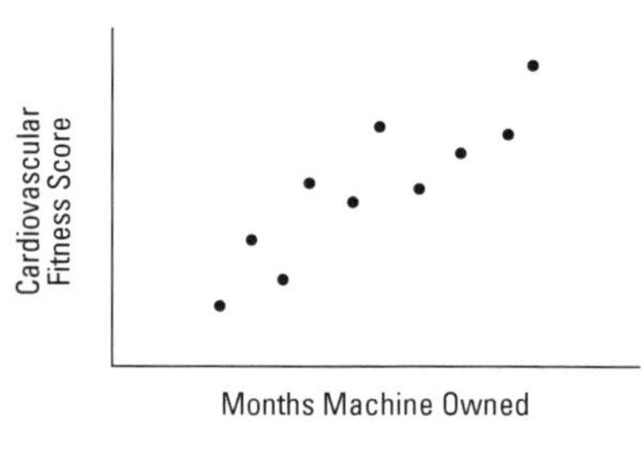

If two variables are positively correlated, as the value of one increases, so does the value of the other. If they are negatively (or inversely) correlated, as the value of one increases, the value of the other decreases.

A third possibility remains: that as the value of one variable increases, the value of the other neither increases nor decreases. Figure 8-3 is a scatter plot of months the exercise machine has been owned (horizontal axis) by the person's height (vertical axis). No line trend can be seen in the plot. New owners of exercise machines may be short or tall, and the same is true of people who have had their machines longer. These two variables appear to be uncorrelated.

Figure 8-3 An example of uncorrelated data.

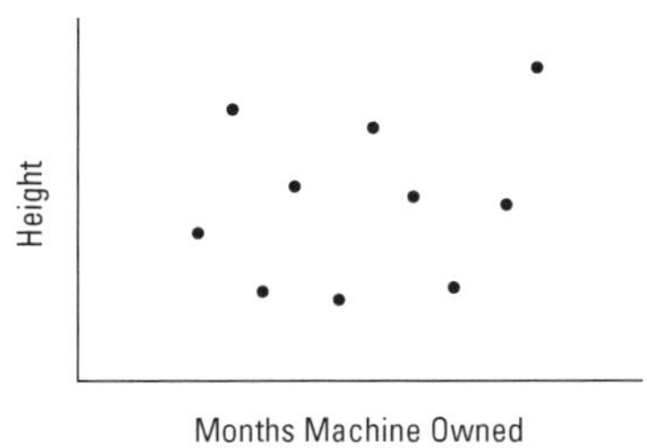

You can go even further in expressing the relationship between variables. Compare the two scatter plots in Figure 8-4. Both plots show a positive correlation because, as the values on one axis increase, so do the values on the other. But the data points in Figure 8-4(b) are more closely packed than the data points in Figure 8-4(a), which are more spread out. If a line were drawn through the middle of the trend, the points in Figure 8-4(b) would be closer to the line than the points in Figure 8-4(a). In addition to direction (positive or negative), correlations also can have strength, which is a reflection of the closeness of the data points to a perfect line. Figure 8-4(b) shows a stronger correlation than Figure 8-4(a).

Figure 8-4 (a) Weak and (b) strong correlations.

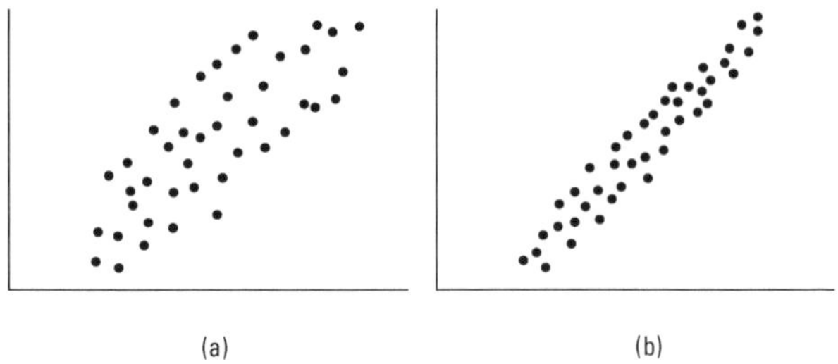

Pearson's product moment coefficient ***(r),*** commonly referred to as the correlation coefficient, is a quantitative measure of correlation between two interval-level variables. The coefficient r can take values from –1.0 to 1.0. The sign of r indicates whether the correlation is positive or negative. The magnitude (absolute value) of r indicates the strength of the correlation, or how close the array of data points is to a straight line.

Two computing formulas for r are

$$r = \frac{\sum xy - \frac{1}{n}\left(\sum x\right)\left(\sum y\right)}{(n-1)s_x s_y}$$

and

$$r = \frac{\sum\left(\frac{x_i - \overline{x}}{s_x}\right)\left(\frac{y_i - \overline{y}}{s_y}\right)}{n-1}$$

where Σxy is the sum of the xy cross-products (each x multiplied by its paired y), n is the size of the sample (the number of data pairs), Σx and Σy are the sums of the x and y values, s_x and s_y are the sample standard deviations of x and y, and $\overline{x}$ and $\overline{y}$ are the means.

Example 1: Use Table 8-2 to compute r for the relationship between months of exercise-machine ownership and hours of exercise per week. The first step is to compute the components required in the main formula. Let x be months of ownership and y be hours of exercise, although you could also do the reverse.

Table 8-2 Determining the Correlation Coefficient for Table 8-1

x	y	xy	x^2	y^2
5	5	25	25	25
10	2	20	100	4
4	8	32	16	64
8	3	24	64	9
2	8	16	4	64
7	5	35	49	25
9	5	45	81	25
6	7	42	36	49
1	10	10	1	100
12	3	36	144	9
$\Sigma x = 64$	$\Sigma y = 56$	$\Sigma xy = 285$	$\Sigma x^2 = 520$	$\Sigma y^2 = 374$

$$\left(\sum x\right)^2 = 64^2 = 4{,}096$$

$$\left(\sum y\right)^2 = 56^2 = 3{,}136$$

Now, compute the sample standard deviations for x and y using the formula from Chapter 3:

$$\begin{aligned} s_x &= \sqrt{\frac{\sum x^2 - \frac{\left(\sum x\right)^2}{n}}{n-1}} \\ &= \sqrt{\frac{520 - \frac{4{,}096}{10}}{10-1}} \\ &= \sqrt{\frac{110.4}{9}} \\ &= \sqrt{12.267} \\ &= 3.502 \end{aligned}$$

$$s_y = \sqrt{\frac{374 - \frac{3,136}{10}}{10 - 1}}$$
$$= \sqrt{\frac{60.4}{9}}$$
$$= \sqrt{6.711}$$
$$= 2.591$$

Finally, r may be computed:

$$r_{x,y} = \frac{\sum xy - \frac{1}{n}\left(\sum x\right)\left(\sum y\right)}{(n-1)s_x s_y}$$
$$= \frac{285 - \frac{1}{10}(64)(56)}{(10-1)(3.502)(2.591)}$$
$$= \frac{-73.4}{81.663}$$
$$= -0.899$$

A correlation of $r = -0.899$ is almost as strong as the maximum negative correlation of -1.0, reflecting the fact that your data points fall relatively close to a straight line.

Finding the significance of *r*

You might want to know how significant an r of -0.899 is. The formula to test the null hypothesis that R (the population correlation) $= 0$ is

$$t = \frac{r\sqrt{(n-2)}}{\sqrt{(1-r^2)}}$$

where r is the sample correlation coefficient, and n is the size of the sample (the number of data pairs). The probability of t may be looked up in Table 3 (in Appendix B) using $n - 2$ degrees of freedom.

$$t = \frac{-0.899\sqrt{(10-2)}}{\sqrt{1-(-0.899)^2}} = \frac{-2.543}{0.438} = -5.807$$

The probability of obtaining a *t* of –5.807 with 8 *df* (drop the sign when looking up the value of *t*) is lower than the lowest listed probability of 0.0005. If the correlation between months of exercise-machine ownership and hours of exercise per week were actually 0, you would expect an *r* of –0.899 or lower in fewer than one out of a thousand random samples.

To evaluate a correlation coefficient, first determine its significance. If the probability that the coefficient resulted from chance is not acceptably low, the analysis should end there; neither the coefficient's sign nor its magnitude may reflect anything other than sampling error. If the coefficient is statistically significant, the sign should give an indication of the direction of the relationship, and the magnitude indicates its strength. Remember, however, that all statistics become significant with a high enough *n*.

Even if it's statistically significant, whether a correlation of a given magnitude is substantively or practically significant depends greatly on the phenomenon being studied. Generally, correlations tend to be higher in the physical sciences, where relationships between variables often obey uniform laws, and lower in the social sciences, where relationships may be harder to predict. A correlation of 0.4 between a pair of sociological variables may be more meaningful than a correlation of 0.7 between two variables in physics.

Bear in mind also that the correlation coefficient measures only straight-line relationships. Not all relationships between variables trace a straight line. Figure 8-5 shows a curvilinear relationship such that values of *y* increase along with values of *x* up to a point, then decrease with higher values of *x*. The correlation coefficient for this plot is 0, the same as for the plot in Figure 8-3. This plot, however, shows a relationship that Figure 8-3 does not.

Correlation does not imply causation. The fact that Variable A and Variable B are correlated does not necessarily mean that A caused B or that B caused A (though either may be true). If you were to examine a database of demographic information, for example, you would find that the number of churches in a city is correlated with the number of violent crimes in the city. The reason is not that church attendance causes crime, but that these two variables both increase as a function of a third variable: population.

Figure 8-5 Data that can cause trouble with correlation analysis.

Also note that the scales used to measure two variables have no effect on their correlation. If you had converted hours of exercise per week to minutes per week, and/or months machine owned to days owned, the correlation coefficient would have been the same.

The coefficient of determination

The square of the correlation coefficient r is called the **coefficient of determination** and can be used as a measure of the proportion of variability that two variables share, or how much one can be "explained" by the other. The coefficient of determination for this example is $(-0.899)^2 = 0.808$. Approximately 80 percent of the variability of each variable is shared with the other variable.

Simple Linear Regression

To describe the linear association between quantitative variables, a statistical procedure called **regression** often is used to construct a model. Regression is used to assess the contribution of one or more "explanatory" variables (called **independent** variables) to one "response" (or **dependent**) variable. It also can be used to predict the value of one variable based on the values of others. When there is only one independent variable and when the relationship can be expressed as a straight line, the procedure is called simple linear regression.

Any straight line in two-dimensional space can be represented by this equation:

$$y = a + bx$$

where y is the variable on the vertical axis, x is the variable on the horizontal axis, a is the y-value where the line crosses the vertical axis (often called the **intercept**), and b is the amount of change in y corresponding to a one-unit increase in x (often called the **slope**). Figure 8-6 gives an example.

Figure 8-6 A straight line.

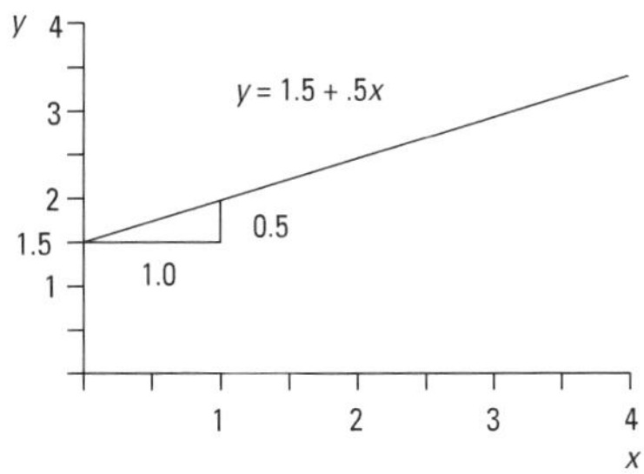

Returning to the exercise example, you observed that the scatter plot of points in Figure 8-1 resembles a line. The regression procedure fits the best possible straight line to an array of data points. If no single line can be drawn such that all the points fall on it, what is the "best" line? Statisticians use the line that minimizes the sum of squared deviations from each data point to the line, a fact that will become clearer after you compute the line for the example.

Regression is an inferential procedure, meaning that it can be used to draw conclusions about populations based on samples randomly drawn from those populations. Suppose that your ten exercise-machine owners were randomly selected to represent the population of all exercise-machine owners. In order to use this sample to make educated guesses about the relationship between the two variables (months of machine ownership and time spent exercising) in the population, you need to rewrite the equation above to reflect the fact that you will be estimating population parameters:

$$y = \beta_0 + \beta_1 x$$

All you have done is replace the intercept (*a*) with β_0 and the slope (*b*) with β_1. The formula to compute the parameter estimate $\hat{\beta}_1$ is

$$\hat{\beta}_1 = \frac{S_{xy}}{S_{xx}}$$

where $S_{xy} = \sum xy - \frac{(\sum x)(\sum y)}{n}$

and $s_{xx} = \sum x^2 - \frac{(\sum x)^2}{n}$

The formula to compute the parameter estimate $\hat{\beta}_0$ is

$$\hat{\beta}_0 = \bar{y} - \hat{\beta}_1\bar{x}$$

where $\bar{y}$ and $\bar{x}$ are the two sample means.

You already have computed the quantities that you need to substitute into these formulas for the exercise example—except for the $\bar{x}$ mean of *x*, which is $\frac{64}{10} = 6.4$, and the $\bar{y}$ mean of *y*, which is $\frac{56}{10} = 5.6$. First, compute the estimate of the slope:

$$S_{xy} = 285 - \frac{(64)(56)}{10} = -73.4$$

$$S_{xx} = 520 - \frac{4{,}096}{10} = 110.4$$

$$\hat{\beta}_1 = \frac{-73.4}{110.4} = -0.665$$

Now the intercept may be computed:

$$\begin{aligned}\hat{\beta}_0 &= \bar{y} - \hat{\beta}_1\bar{x} \\ &= 5.6 - (-0.665(6.4)) \\ &= 5.6 - (-4.256) \\ &= 9.856\end{aligned}$$

So, the regression equation for the example is $y = 9.856 - 0.665x$. When you plot this line over the data points, the result looks like that shown in Figure 8-7.

Figure 8-7 Illustration of residuals.

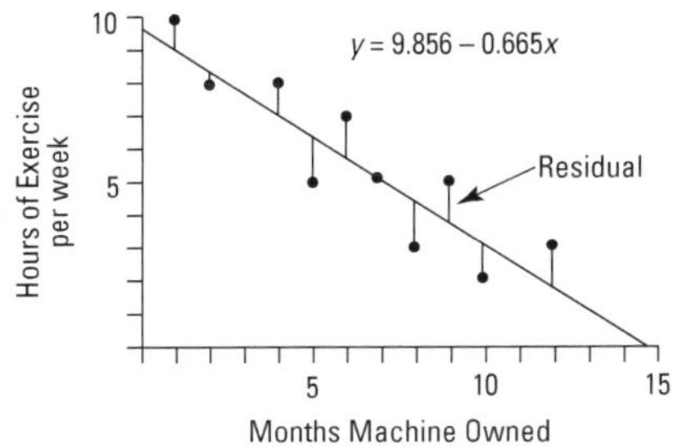

The vertical distance from each data point to the regression line is the error, or **residual,** of the line's accuracy in estimating that point. Some points have positive residuals (they lie above the line); some have negative ones (they lie below it). If all the points fell on the line, there would be no error and no residuals. The mean of the sample residuals is always 0 because the regression line is always drawn such that half of the error is above it and half below it. The equations that you used to estimate the intercept and slope determine a line of "best fit" by minimizing the sum of the squared residuals. This method of regression is called **least squares.**

Because regression estimates usually contain some error (that is, all points do not fall on the line), an error term (ε, the Greek letter epsilon) is usually added to the end of the equation:

$$y = \beta_0 + \beta_1 x + \varepsilon$$

The estimate of the slope β_1 for the exercise example was –0.665. The slope is negative because the line slants down from left to right, as it must for two variables that are negatively correlated, reflecting that one variable decreases as the other increases. When the correlation is positive, β_1 is positive, and the line slants up from left to right.

Confidence interval for the slope

Example 2: What if the slope is 0, as in Figure 8-8? That means that y has no linear dependence on x, or that knowing x does not contribute anything to your ability to predict y.

Figure 8-8 An example of uncorrelated data, so the slope is zero.

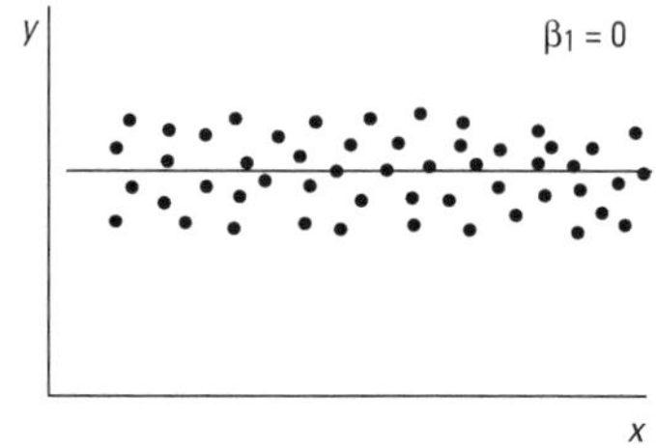

It is often useful to compute a confidence interval for a regression slope. If it contained 0, you would be unable to conclude that x and y are related. The formula to compute a confidence interval for β_1 is

$$(a,b) = \hat{\beta}_1 \pm t_{\alpha/2,n-2} \cdot \frac{s}{\sqrt{S_{xx}}}$$

where $s = \sqrt{\dfrac{\sum(y-\hat{y})^2}{n-2}}$

and $S_{xx} = \sum x^2 - \dfrac{\left(\sum x\right)^2}{n}$

and where $\sum(y-\hat{y})^2$ is the sum of the squared residuals, $t_{\alpha/2,n-2}$ is the critical value from the t-table corresponding to half the desired alpha level at $n - 2$ degrees of freedom, and n is the size of the sample (the number of data pairs). The test for this example will use an alpha of 0.05. Table 3 (in Appendix B) shows that $t_{.025,8} = 2.306$.

Compute the quantity $\sum(y-\hat{y})^2$ by subtracting each predicted y-value $(\hat{y})$ from each actual y-value, squaring it, and summing the squares (see Table 8-3). The predicted y-value $(\hat{y})$ is the y-value that would be predicted from each given x, using the formula $y = 9.856 - 0.665x$.

Table 8-3 Determining the Residuals for the Data in Table 8-1

x	*y*		$\hat{y}$		*residual*	*residual*^{2}
5	5	–	6.531	=	–1.531	2.344
10	2	–	3.206	=	–1.206	1.454
4	8	–	7.196	=	0.804	0.646
8	3	–	4.536	=	–1.536	2.359
2	8	–	8.526	=	–0.526	0.277
7	5	–	5.201	=	–0.201	0.040
9	5	–	3.871	=	1.129	1.275
6	7	–	5.866	=	1.134	1.286
1	10	–	9.191	=	0.809	0.654
12	3	–	1.876	=	1.124	1.263
					0	11.600

Now, compute *s:*

$$s = \sqrt{\frac{11.600}{10-2}} = \sqrt{1.45} = 1.204$$

You have already determined that S_{xx} = 110.4; you can proceed to the main formula:

$$\begin{aligned}(a,b) &= -0.665 \pm 2.306\frac{1.204}{\sqrt{110.4}}\\ &= -0.665 \pm 2.306\frac{1.204}{10.507}\\ &= -0.665 \pm 2.306(0.115)\\ &= -0.665 \pm 0.264\\ &= (-0.929, -0.401)\end{aligned}$$

You can be 95 percent confident that the population parameter β_1 (the slope) is no lower than –0.929 and no higher than –0.401. Because this interval does not contain 0, you would be able to reject the null hypothesis that $\beta_1 = 0$ and conclude that these two variables are indeed related in the population.

Confidence interval for prediction

You have learned that you could predict a *y*-value from a given *x*-value. Because there is some error associated with your prediction, however, you might want to produce a confidence interval rather than a simple point estimate. The formula for a prediction interval for *y* for a given *x* is

$$(a,b)=\hat{y}\pm t_{\alpha/2,n-2}(s)\sqrt{1+\frac{1}{n}+\frac{(x-\bar{x})^2}{S_{xx}}}$$

where $S_{xx}=\sum x^2-\frac{\left(\sum x\right)^2}{n}$

and $s=\sqrt{\frac{\sum(y-\hat{y})^2}{n-2}}$

and where $\hat{y}$ is the *y*-value predicted for *x* using the regression equation, $t_{\alpha/2,n-2}$ is the critical value from the *t*-table corresponding to half the desired alpha level at $n-2$ degrees of freedom, and *n* is the size of the sample (the number of data pairs).

Example 3: What is a 90 percent confidence interval for the number of hours spent exercising per week if the exercise machine is owned 11 months?

The first step is to use the original regression equation to compute a point estimate for *y:*

$$\begin{aligned} y &= 9.856-0.665(11) \\ &= 9.856-7.315 \\ &= 2.541 \end{aligned}$$

For a 90 percent confidence interval, you need to use $t_{.05,8}$, which Table 3 (in Appendix B) shows to be 1.860. You have already computed the remaining quantities, so you can proceed with the formula:

$$\begin{aligned}(a,b) &= 2.541 \pm 1.860(1.204)\sqrt{1+\frac{1}{10}+\frac{(11-6.4)^2}{110.40}} \\ &= 2.541 \pm 1.860(1.204)\sqrt{1+0.10+\frac{21.16}{110.40}} \\ &= 2.541 \pm 1.860(1.204)\sqrt{1.292} \\ &= 2.541 \pm 2.545 \\ &= (-0.004, 5.086)\end{aligned}$$

You can be 90 percent confident that the population mean for the number of hours spent exercising per week when x (number of weeks machine owned) = 11 is between about 0 and 5.

Assumptions and cautions

The use of regression for parametric inference assumes that the errors (ε) are (1) independent of each other and (2) normally distributed with the same variance for each level of the independent variable. Figure 8-9 shows a violation of the second assumption. The errors (residuals) are greater for higher values of x than for lower values.

Figure 8-9 Data with increasing variance as x increases.

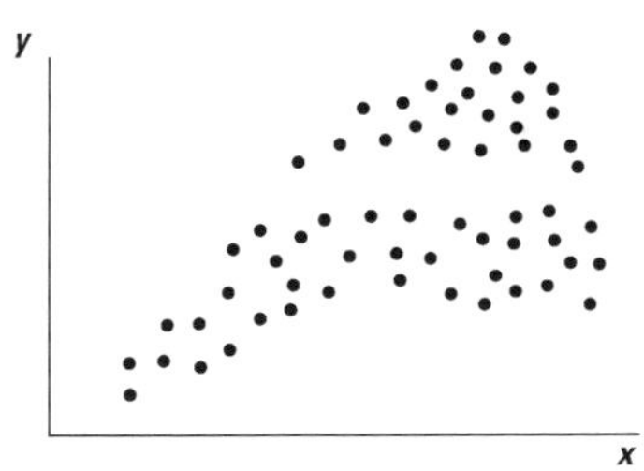

Least squares regression is sensitive to outliers, or data points that fall far from most other points. If you were to add the single data point $x = 15$, $y = 12$ to the exercise data, the regression line would change to the dotted line shown in Figure 8-10. You need to be wary of outliers because they can influence the regression equation greatly.

Figure 8-10 Least squares regression is sensitive to outliers.

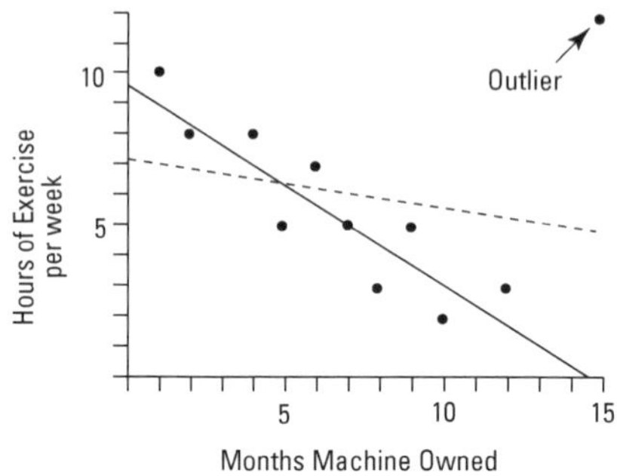

It can be dangerous to extrapolate in regression—to predict values beyond the range of your data set. The regression model assumes that the straight line extends to infinity in both directions, which often is not true. According to the regression equation for the example, people who have owned their exercise machines longer than around 15 months do not exercise at all. It is more likely, however, that "hours of exercise" reaches some minimum threshold and then declines only gradually, if at all (see Figure 8-11).

Figure 8-11 Extrapolation beyond the data is dangerous.

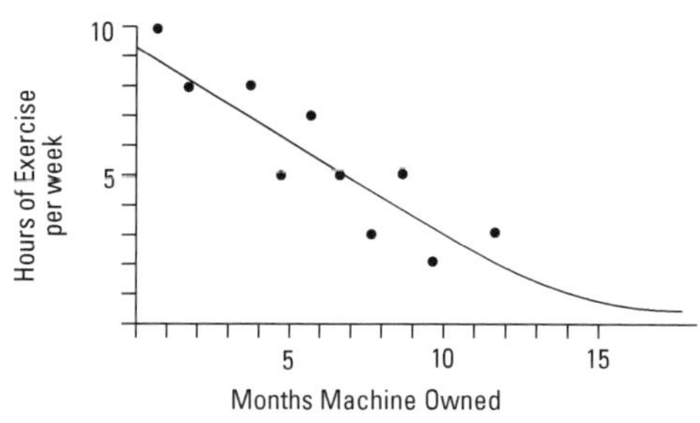

Chi-Square (χ^2)

The statistical procedures that we have reviewed thus far are appropriate only for numerical variables. The **chi-square** (χ^2) test can be used to evaluate a relationship between two categorical variables. It is one example of a **nonparametric test.** Nonparametric tests are used when assumptions about normal distribution in the population cannot be met. These tests are less powerful than parametric tests.

Suppose that 125 children are shown three television commercials for breakfast cereal and are asked to pick which they liked best. The results are shown in Table 8-4.

You would like to know if the choice of favorite commercial was related to whether the child was a boy or a girl or if these two variables are independent. The totals in the margins will allow you to determine the overall probability of (1) liking commercial A, B, or C, regardless of gender, and (2) being either a boy or a girl, regardless of favorite commercial. If the two variables are independent, then you should be able to use these probabilities to predict approximately how many children should be in each cell. If the actual count is very different from the count that you would expect if the probabilities are independent, the two variables must be related.

Table 8-4 Commercial Preference for Boys and Girls

	A	*B*	*C*	*Totals*
Boys	30	29	16	75
Girls	12	33	5	50
Totals	42	62	21	125

Consider the upper-right cell of the table. The overall probability of a child in the sample being a boy is $75 \div 125 = 0.6$. The overall probability of liking Commercial A is $42 \div 125 = 0.336$. The multiplication rule (covered in Chapter 4) states that the probability of both of two independent events occurring is the product of their two probabilities. Therefore, the probability of a child both being a boy and liking Commercial A is $0.6 \times 0.336 = 0.202$. The expected number of children in this cell, then, is $0.202 \times 125 = 25.2$.

There is a faster way of computing the expected count for each cell: Multiply the row total by the column total and divide by *n*. The expected count for the first cell is, therefore, $(75 \times 42) \div 125 = 25.2$. If you perform this operation for each cell, you get the expected counts (in parentheses) shown in Table 8-5.

Table 8-5 Chi-Square Results for Table 8-4

	A	*B*	*C*	*Totals*
Boys	30 (25.2)	29 (37.2)	16 (12.6)	75
Girls	12 (16.8)	33 (24.8)	5 (8.4)	50
Totals	42	62	21	125

Note that the expected counts properly add up to the row and column totals. You are now ready for the formula for χ^2, which compares each cell's actual count to its expected count:

$$\chi^2 = \sum \frac{(\text{observed} - \text{expected})^2}{\text{expected}}$$

The formula describes an operation that is performed on each cell and which yields a number. When all the numbers are summed, the result is χ^2. Now, compute it for the six cells in the example:

$$\begin{aligned}\chi^2 &= \frac{(30-25.2)^2}{25.2} + \frac{(29-37.2)^2}{37.2} + \frac{(16-12.6)^2}{12.6} + \\ &\quad \frac{(12-16.8)^2}{16.8} + \frac{(33-24.8)^2}{24.8} + \frac{(5-8.4)^2}{8.4} \\ &= 0.914 + 1.808 + 0.917 + 1.371 + 2.711 + 1.376 \\ &= 9.098\end{aligned}$$

The larger χ^2, the more likely that the variables are related; note that the cells that contribute the most to the resulting statistic are those in which the expected count is very different from the actual count.

Chi-square has a probability distribution, the critical values for which are listed in Table 4 in Appendix B. As with the *t*-distribution, χ^2 has a degrees-of-freedom parameter, the formula for which is

$$(\text{number of rows} - 1) \times (\text{number of columns} - 1)$$

or in your example:

$$(2 - 1) \times (3 - 1) = 1 \times 2 = 2$$

In Table 4 in Appendix B, a chi-square of 9.097 with two degrees of freedom falls between the commonly used significance levels of 0.05 and 0.01. If you had specified an alpha of 0.05 for the test, you could, therefore, reject the null hypothesis that gender and favorite commercial are independent. At $a = 0.01$, however, you could not reject the null hypothesis.

The χ^2 test does not allow you to conclude anything more specific than that there is some relationship in your sample between gender and commercial liked (at $\alpha = 0.05$). Examining the observed versus expected counts in each cell might give you a clue as to the nature of the relationship and which levels of the variables are involved. For example, Commercial B appears to have been liked more by girls than boys. But χ^2 tests only the very general null hypothesis that the two variables are independent.

Sometimes a chi-square test of homogeneity of populations is used. It is very similar to the test for independence. In fact the mechanics of these tests are identical. The real difference is in the design of the study and the sampling method.

Chapter Check-Out

Q&A

For the problems 1 through 3, consider this data on the dose of Medicine X administered and the length of time of continued illness.

Dose	0	1	2	3	4	5	6
Time to cure (days)	15.4	12.2	13.7	9.2	9.9	6.1	4.1

1. a. Make a scatter plot of the data. Is this negative or positive correlation?
   b. Calculate the correlation coefficient.
   c. Find the significance of the correlation coefficient.
2. a. Apply linear regression to the data to find the slope and intercept.
   b. Calculate the 95 percent confidence interval for the slope of your regression.
3. a. Predict the result if the dose was 1.5, and give the 90 percent confidence interval for your prediction.
   b. Predict the result if the dose was 8, and give the 90 percent confidence interval.

Answers 1a. negative **b.** –0.9511 **c.** $t = -6.885$, $p < 0.0005$ **2a.** $\beta_1 = -1.78$; $\beta_0 = 15.4321$ **b.** -1.78 ± 0.66 **3a.** 12.76 ± 3.05 **b.** 1.19 ± 3.94

Critical Thinking

1. Imagine that the experiment in problems 1 through 3 is carried further, and the results are as follows:

Dose	7	8	9	10	11
Time to cure (days)	4.0	3.9	4.2	3.8	4.5

Add this data to your scatter plot and comment on what is happening. What about your prediction in problem 3b?

REVIEW

Use this Review to practice what you've learned in this book. After you work through the review questions, you're well on your way to achieving your goal of understanding basic statistical methods.

Chapter 2

1. What is wrong with this pie chart?

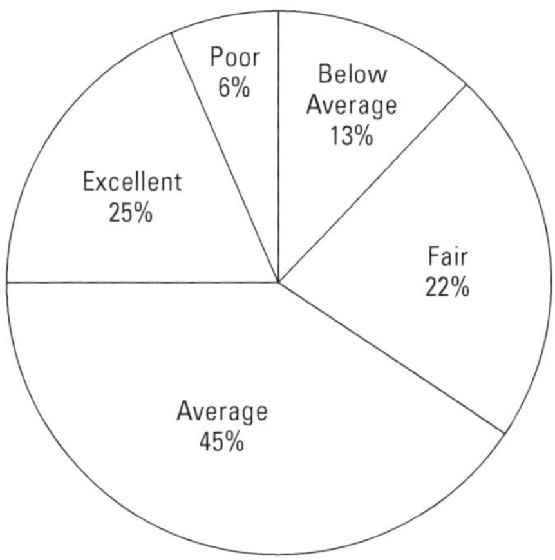

2. Study the following frequency histogram of the inventory at a hat store. The total number of hats is 1,000. Then answer these questions:

a. How many hats are blue?

b. How many hats are white or black?

c. If a hat is chosen randomly, what is the chance that it is multicolored?

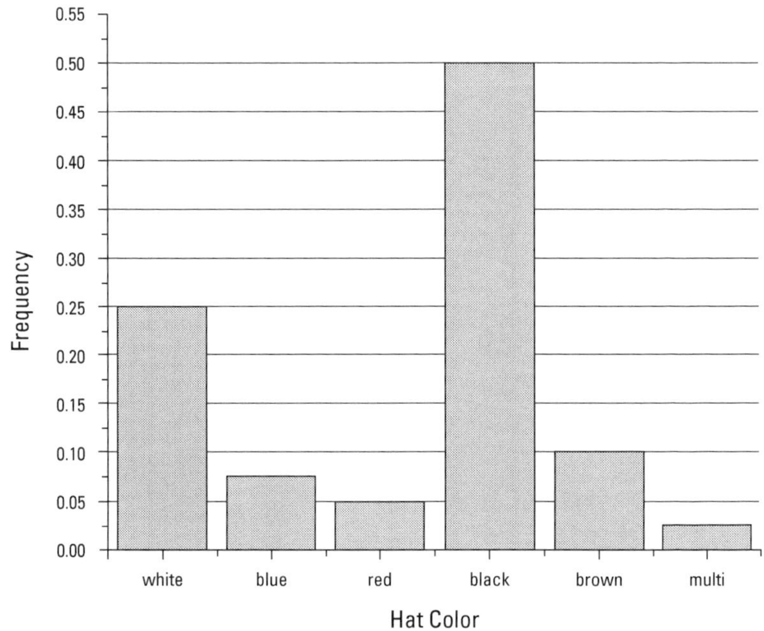

3. True or False: *Median* and *mean* are two terms for the same quantity.

4. True or False: *Variance* and *standard deviation* are two terms for the same quantity.

Chapter 4

5. True or False: The binomial distribution applies only to events that have two outcomes: "success" or "failure."

6. A fair die, with six sides, is tossed twice. The chance of rolling two sixes is

a. $\frac{2}{6}$

b. $\frac{1}{6}$

c. $\frac{1}{12}$

d. $\frac{1}{36}$

e. $\frac{1}{4}$

7. Two fair dice, with six sides each, are rolled. The probability that the sum of the dice equals 12 is

a. 0

b. 1

c. $\frac{1}{6}$

d. $\frac{1}{12}$

e. $\frac{1}{36}$

8. Two fair dice, with six sides each, are rolled. The probability that the sum of the dice equals 7 is

a. 0

b. 1

c. $\frac{1}{6}$

d. $\frac{1}{12}$

e. $\frac{1}{36}$

9. A population with a normal distribution has a mean of 1 and a variance of 4. The best you can say about the probability of randomly choosing a value greater than 7 from this sample is that it is

a. 0

b. less than 0.15 percent

c. less than 0.3 percent

d. less than 32 percent

e. less than 16 percent

10. A slot machine has four windows, and each can contain one of eight different symbols. To win the jackpot, all four windows must contain the same symbol. The chance of a jackpot is

a. 0

b. $\frac{1}{8}$

c. $\left(\frac{1}{8}\right)^3$

d. $\frac{1}{4}$

e. 1

Chapter 5

11. True or False: In a normally distributed population, the standard deviation may never be larger than the mean.

12. True or False: For all practical purposes, μ and $\bar{x}$ are the same thing.

13. Which of the following is not a correct null hypothesis H_0?

a. $\mu = 1$

b. $\mu_1 = \mu_2$

c. $\pi = 0.5$

d. $\pi_1 = \pi_2$

e. $\bar{x} = 1$

14. Which normal curve pictured below has the greatest standard deviation?

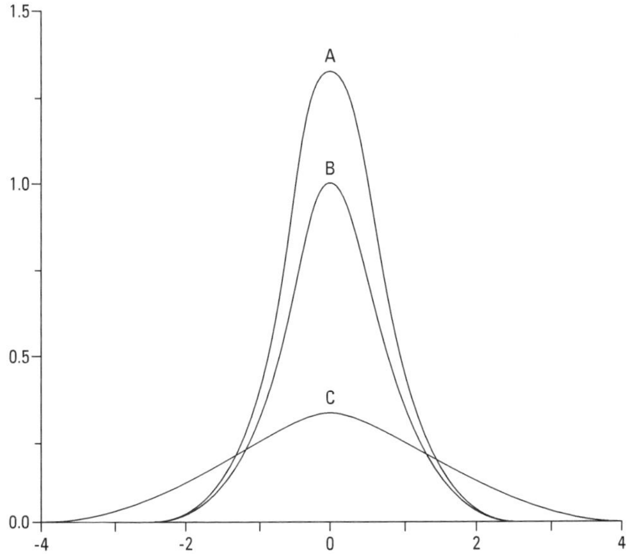

Chapter 6

15. True or False: If two variables are correlated, then they must be causally related.

16. Rejecting the null hypothesis when it is actually true is

a. no error
b. a Type I error
c. a Type II error
d. neither a Type I nor a Type II error
e. impossible

Chapter 7

17. A machine for filling beer bottles fills the bottles so that the distribution has a mean of 12 ounces of beer in each bottle, with a standard deviation of 0.05 ounce. What is the probability that a randomly selected bottle contains at least 11.9 ounces?

18. Suppose that there are only two candidates (A and B) for president and that in a particular state 5,000,000 votes will be cast.

- **a.** What percentage of the total votes should a TV station count, to be sure to within ±1 percent that candidate A has won, at a 95 percent confidence level?
- **b.** The first 20,000 votes are rapidly counted. Candidate A has received 52 percent, and candidate B has received 48 percent of the vote. What is the width of the 99.9 percent confidence interval on candidate A's result?
- **c.** Comment on the methods of projection in part **a.** Is it a good idea to use the first votes cast to project the winner?

19. At a major airport, a researcher studies the average length of delays (excluding flights that leave on time). His data is shown below, for ten randomly chosen days.

Day	1	2	3	4	5	6	7	8	9	10
Average delay (hours)	2.0	1.8	0.32	0.41	2.3	0.72	0.22	0.28	0.57	3.2

- **a.** Calculate the mean and standard deviation of the delays, as well as the median.
- **b.** What is the 95 percent confidence interval on the average length of delay?
- **c.** Compare the mean and median from part **a.** What is the p value for a delay of the median size?
- **d.** Do you think this population is normally distributed?

Chapter 8

20. True or False: If the correlation coefficient for a set of data is zero, then there is no relationship between the dependent and independent variables.

21. True or False: The χ^2 test tells you only if two variables are independent.

22. A microbiologist is testing a new antibiotic she developed against a common one. She applies each drug to different plates of bacterial cultures and measures the fraction of bacteria that dies after one day. Her results are given in the following table.

Dose of Drug (mg)	0	0.5	1.0	1.5	2.0	2.5	3.0
New drug (% dead)	2.0	12.3	27.1	32.2	48.9	56.6	60.6
Old drug (% dead)	3.1	6.0	16.5	12.3	25.9	29.0	42.5

Using the information in the preceding table, answer these questions:

a. Plot the results for each drug and calculate the coefficient of correlation for each drug.

b. What is the significance level of each of these correlations?

c. Calculate the slope for each plot.

d. Do the 95 percent confidence intervals for the two slopes overlap?

23. Consider the following data on the color of a dog and the average annual income of the dog's owner. Compute a χ^2 value for this data and determine the p-level to which these variables are related.

	\$0–\$25,000	***\$25,001–\$50,000***	***\$50,001–\$75,000***	***\$75,000+***	***Totals***
White	10	12	17	23	62
Black	15	28	13	11	67
Orange/golden	7	9	11	15	42
Mixed	24	26	32	42	124
Total	56	75	73	91	295

Answers: 1. More than 100 percent is shown. **2a.** 75 **b.** 750 **c.** 0.025 **3.** False **4.** False **5.** True **6.** d **7.** e **8.** c **9.** b **10.** c **11.** False **12.** False **13.** e **14.** c **15.** False **16.** b **17.** 97.72 percent **18a.** 0.192 percent **b.** 2.3 percent **19a.** 1.18; 1.06; 0.65 **b.** (0.42, 1.94) **c.** $0.05 < p < 0.1$ **d.** No **20.** False **21.** True **22a.** 0.989853; 0.95892 **b.** $p < 0.0005$; $p < 0.0005$ **c.** 20.44; 12.4 **d.** No **23.** $\chi^2 = 17.03406$; $0.025 < p < 0.05$.

RESOURCE CENTER

This Resource Center offers the best resources available in print and online to help you study and review the core concepts of statistics and statistical methods. You can find additional resources, plus study tips and tools to help test your knowledge, at www.cliffsnotes.com.

Books

There are many books available on basic statistical methods. Many use humor to get across the concepts. Although most of these concepts are covered in this book, if you are still struggling, one of these may help.

Excel 2010 For Dummies, by Greg Harvey (Wiley). Once you have your data, how can you analyze it efficiently? After all, if you have more than ten samples, you do not want to be calculating by hand. You need a spreadsheet, and Excel performs good statistical analysis, including all the basic techniques from this book. *Excel 2010 For Dummies* shows you how to use Excel effectively and how to perform the statistical tests you are interested in.

The Cartoon Guide to Statistics, by Larry Gonick and Woollcott Smith (HarperCollins). This book offers much of the same information as *CliffsNotes Statistics Quick Review,* but with cartoons! It may sound silly, but the cartoons really can help to illustrate the concepts.

Chances Are: The Only Statistics Book You'll Ever Need, by Steve Slavin (Madison Books). This book is designed to make statistics accessible to people with little mathematical background. If your math skills are holding you up, check this book out.

How to Lie with Statistics, by Darrell Huff (W. W. Norton & Co.). This is a classic, and any list of statistics books would be incomplete without it. Huff shows how badly analyzed and interpreted data can be used to sell products or push agendas, providing good examples of the misuses and abuses of statistics. Worth a look, to keep from repeating the errors he points out.

Internet

Visit the following Web sites for more information about statistics:

American Statistical Association (www.amstat.org). This site is more interesting than you might think. There is a nice discussion of what statisticians do, career opportunities in statistics, and also some educational resources. The association offers online courses and maintains a list of other Web sites related to statistics.

Concepts and Applications of Inferential Statistics (http://faculty.vassar.edu/~lowry/webtext.html). This site is a complete statistics textbook! If you want to review the concepts in *CliffsNotes Statistics Quick Review* or move beyond into some more advanced concepts, this site is worth a look. The explanations are clear, and examples abound. The site is maintained by Professor Richard Lowry at Vassar College.

Portraits of Statisticians (www.york.ac.uk/depts/maths/histstat/people). If you want to see where all this statistics stuff came from, well, here are pictures or portraits of famous statisticians, past and present. An interesting site put together by the mathematics department at the University of York in the United Kingdom.

Statistical Methods (http://bubl.ac.uk/link/s/statisticalmethods.htm). This is actually a page from the BUBL Catalog on Internet Resources, a larger site maintained by the Centre for Digital Library Research at Strathclyde University in the United Kingdom. On this page, you will find many links to pages dedicated to statistical methods. The BUBL page is well organized, with a description of the page that each link leads to.

Note: The Internet is loaded with statistical information. Most of it, however, is descriptive statistics. These pages include all sorts of government statistics—from the U.S. Bureau of Transportation (www.bts.gov), to the Bureau of Justice Statistics (http://bjs.ojp.usdoj.gov). There is a vast amount of information, to which you could apply the statistical methods you learned with this book.

Next time you're on the Internet, don't forget to drop by www.cliffsnotes.com. We created an online Resource Center that you can use today, tomorrow, and beyond.

GLOSSARY

addition rule For mutually exclusive random events, the chance of at least one of them occurring is the sum of their individual probabilities.

alternative hypothesis A research hypothesis; the hypothesis that is supported if the null hypothesis is rejected.

bar chart A graphic that displays how data fall into different categories or groups.

bell-shaped curve Symmetrical, single-peaked frequency distribution. Also called the normal curve or Gaussian curve.

bias The systematic underestimation or overestimation of a population parameter, because of the method used to select the sample, or the method used to obtain the information.

bimodal Curve with two equal scores of highest frequency.

binomial An event with only two possible outcomes.

binomial probability distribution For binomial events, the frequency of the number of favorable outcomes. For a large number of trials, the binomial distribution approaches the normal distribution.

bivariate Involving two variables, especially, when attempting to show a correlation between two variables, the analysis is said to be bivariate.

box plot (box-and-whiskers) A graphic display of data indicating symmetry and central tendency.

central limit theorem A rule that states that the sampling distribution of means from any population will be normal for large sample n.

chi-square A probability distribution used to test the independence of two categorical variables.

class frequency The number of observations that fall into each class interval.

class intervals Categories or groups contained in frequency graphics.

coefficient of determination A measure of the proportion of each other's variability that two variables share.

confidence interval The range of values that a population parameter could take at a given level of significance.

confidence level A selected percentage of confidence for a confidence interval.

continuous variable A variable that can be measured with whole numbers and fractional (or decimal) parts thereof.

correlated Two (or more) quantities that change together in a consistent manner. Thus, if the value of one variable is known, the other can be predicted from their relationship.

correlation coefficient A measure of the degree to which two variables are linearly related.

critical value The value of a computed statistic used as a threshold to decide whether the null hypothesis will be rejected.

data Numerical information about variables; the measurements or observations to be analyzed with statistical methods.

degrees of freedom A parameter used to help select the critical value in some probability distributions.

dependent events Events such that the outcome of one has an effect on the probability of the outcome of the other.

dependent variable A variable that is caused or influenced by another.

descriptive statistics Numerical data that describe phenomena.

deviation The distance of a value in a population (or sample) from the mean value of the population (or sample).

directional test A test of the prediction that one value is higher than another; also called a one-tailed test.

discrete variable A variable that can be measured by means of whole numbers only; or one that assumes only a certain set of definite values, and no others.

disjoint occurrence Both outcomes unable to happen at the same time.

distribution A collection of measurements; how scores tend to be dispersed about a measurement scale.

dot plot A graphic that displays the variability in a small set of measures.

double-counting A mistake encountered in calculating the probability of at least one of several events occurring, when the events are not mutually exclusive. In this case, the addition rule does not apply.

empirical rule A rule that describes the percentage of values that fall within intervals for a normal distribution. Sometimes called the 68-95-99.7 rule.

frequency distribution The frequency of occurrence of the values of a variable. For each possible value of the variable, there is an associated frequency with which the variable assumes that value.

frequency histogram A graphic that displays how many measures fall into different classes, giving the frequency at which each category is observed.

frequency polygon A graphic presentation of frequency of a phenomenon that typically uses straight lines and points.

grouped data Data that has been sorted into intervals, usually in order to construct a frequency histogram.

grouped measures A set of values that belong to the same class.

histogram A graphic presentation of frequency of a numerical variable.

independent events Events such that the outcome of one has no effect on the probability of the outcome of the other.

independent variable A variable that causes or influences another variable.

inference A conclusion about a population parameter based upon analysis of a sample statistic. Inferences are always stated with a confidence or significance level.

intercept The value of y at which a line crosses the vertical axis.

interquartile range The difference between the upper quartile (75th percentile) and the lower quartile (25th percentile).

interval A scale using numbers to rank order; its intervals are equal but with an arbitrary 0 point.

joint occurrence Both outcomes happening simultaneously; $P(A \cap B)$.

least squares Any line- or curve-fitting model that minimizes the squared distance of data points to the line.

lower quartile (Q_1), the 25th percentile of a set of measures.

mean The sum of the measures in a distribution divided by the number of measures; the average.

measures of central tendency Descriptive measures that indicate the center of a set of values—for example, mean, median, and mode.

measures of variation Descriptive measures that indicate the dispersion of a set of values—for example, variance, standard deviation, and standard error of the mean.

median The middle measure in an ordered distribution.

middle quartile (Q_2), the 50th percentile of a set of measures; the median.

mode Most frequent measure in a distribution; the high point on a frequency distribution.

mound-shaped curve See *bell-shaped curve.*

multiplication rule The probability of two or more independent (hence, not mutually exclusive) events all occurring is the product of their individual probabilities.

mutually exclusive Events such that the occurrence of one precludes the occurrence of the other.

negative relationship A relationship between two variables such that when one increases, the other decreases.

negatively skewed curve A probability or frequency distribution that is not normal but rather is shifted such that the mean is less than the mode.

nominal A scale using numbers, symbols, or names to designate different subclasses.

nondirectional test A test of the prediction that two values are equal or a test that they are not equal; a two-tailed test.

nonparametric test A statistical test used when assumptions about normal distribution in the population cannot be met, or when the level of measurement is ordinal or less. For example, the chi-square test.

normal distribution A specific continuous probability distribution with a graph that is a smooth bell-shaped curve symmetrical about the mean.

null hypothesis The reverse of the research hypothesis. The null hypothesis is directly tested by statistical analysis so that it is either rejected or not rejected, with a significance level. If the null hypothesis is rejected, the alternative hypothesis is supported.

numerical statistics Statistical parameters presented as numbers (as opposed to pictorial statistics).

ogive A graphic that displays a running total.

one-tailed test A test of the prediction that one value is higher than another.

ordinal A scale using numbers or symbols to rank order; its intervals are unspecified.

outlier A data point that falls far from most other points; a score extremely divergent from the other measures of a set.

parameter A characteristic of a population. The goal of statistical analysis is usually to estimate population parameters, using statistics from a sample of the population.

Pearson's product moment coefficient Identical to the correlation coefficient.

percentile The value in an ordered set of measurements such that *P* percent of the measures lie below that value.

pictorial statistics Statistical parameters that are presented as graphs or charts (as opposed to simply as numbers).

pie chart A graphic that displays parts of the whole, in the form of a circle with its area divided proportionally.

point estimate A number computed from a sample to represent a population parameter.

population A group of phenomena that have something in common. The population is the larger group, whose properties (parameters) are estimated by taking a smaller sample from within the population, and applying statistical analysis to the sample.

positive relationship A relationship between two variables such that when one increases, the other increases, or when one decreases, the other decreases.

positively skewed curve A probability or frequency distribution that is not normal, but rather is shifted such that the mean is greater than the mode.

power The probability that a test will reject the null hypothesis when it is, in fact, false.

probability A quantitative measure of the chances for a particular outcome or outcomes.

probability distribution A smooth curve indicating the frequency distribution for a continuous random variable.

proportion For a binomial random event, the probability of a successful (or favorable) outcome in a single trial.

qualitative variable Phenomenon measured in kind—that is, nonnumerical units. For example, color is a qualitative variable, because it cannot be expressed simply as a number.

quantitative variable Phenomenon measured in amounts—that is, numerical units. For example, length is a quantitative variable.

random An event for which there is no way to know, before it occurs, what the outcome will be. Instead, only the probabilities of each possible outcome can be stated.

random error An error that occurs as a result of sampling variability, through no direct fault of the sampler. It is a reflection of the fact that the sample is smaller than the population; for larger samples, the random error is smaller.

range The difference between the largest and smallest measures of a set.

ratio A scale using numbers to rank order; its intervals are equal, and the scale has an absolute-zero point.

region of acceptance The area of a probability curve in which a computed test statistic will lead to acceptance of the null hypothesis.

region of rejection The area of a probability curve in which a computed test statistic will lead to rejection of the null hypothesis.

regression A statistical procedure used to estimate the linear dependence of one or more independent variables on a dependent variable.

relative frequency The ratio of class frequency to total number of measures.

relative frequency principle of probability If a random event is repeated a large number of times, then the proportion of times that a particular outcome occurs is the probability of that outcome occurring in a single event.

research hypothesis A prediction or expectation to be tested. If the null hypothesis is rejected, then the research hypothesis (also called the alternative hypothesis) is supported.

residual The difference of an actual value y and its predicted value.

sample A group of members of a population selected to represent that population. A sample to which statistical analysis is applied should be randomly drawn from the population, to avoid bias.

sampling distribution The distribution obtained by computing a statistic for all possible samples of the selected size drawn from the same population.

sampling variability The tendency of the same statistic computed from a number of random samples drawn from the same population to differ.

scatter plot A graphic display used to illustrate degree of correlation between two variables.

skewed A distribution displaced at one end of the scale and a tail strung out at the other end.

slope A measure of a line's slant.

standard deviation A measure of data variation; the square root of the variance.

standard error A measure of the random variability of a statistic, such as the mean (i.e., standard error of the mean). The standard error of the mean is equal to the standard deviation divided by the square root of the sample size (n).

standardize To convert to a z-score.

statistic A characteristic of a sample. A statistic is an estimate of a population parameter. For larger samples, the statistic is a better estimate of the parameter.

statistical significance The probability of obtaining a given result by chance. High statistical significance does not necessarily imply importance.

statistics A branch of mathematics that describes and reasons from numerical observations; or descriptive measures of a sample.

stem-and-leaf A graphic display that shows actual scores as well as distribution of classes.

symmetric distribution A probability or frequency distribution that has the property in which the mean, median, and mode are all the same value.

symmetry A shape such that one side is the exact mirror image of the other.

systematic error The consistent underestimation or overestimation of a true value, due to poor sampling technique.

***t*-distribution** A probability distribution often used when the population standard deviation is not known or when the sample size is small.

tabled value The value of a computed statistic used as a threshold to decide whether the null hypothesis will be rejected.

test statistic A computed quantity used to decide hypothesis tests.

two-tailed test A test of the prediction that two values are equal, or a test that they are not equal.

Type I error Rejecting a null hypothesis that is, in fact, true.

Type II error Failing to reject a null hypothesis that is, in fact, false.

upper quartile (Q_3), the 75th percentile of a set of measures.

value A measurement or classification of a variable.

variable An observable characteristic of a phenomenon that can be measured or classified.

variance A measure of data variation; the mean of the squared deviation scores about the means of a distribution.

***z*-score** A unit of measurement obtained by subtracting the mean and dividing by the standard deviation.

Appendix A
COMMON MISTAKES

It may seem that there are many ways to make errors in working a statistics problem. In fact, most errors on statistics exams can be reduced to a short list of common oversights. If you learn to avoid the mistakes listed here, you can greatly reduce your chances of making an error on an exam.

- **Forgetting to convert between standard deviation (σ and s) and variance (σ^2 and s^2):** Some formulas use one; some use the other. Square the standard deviation to get the variance, or take the positive square root of the variance to get the standard deviation.
- **Misstating one-tailed and two-tailed hypotheses:** If the hypothesis predicts simply that one value will be higher than another, it requires a one-tailed test. If, however, it predicts that two values will be different—that is, one value will be either higher *or* lower than another or that they will be equal—then use a two-tailed test. Make sure your null and alternative hypotheses together cover all possibilities—greater than, less than, and equal to.
- **Failing to split the alpha level for two-tailed tests:** If the overall significance level for the test is 0.05, then you must look up the critical (tabled) value for a probability of 0.025. The alpha level is always split when computing confidence intervals.
- **Misreading the standard normal (z) table:** As explained in Chapter 5, all standard normal tables do not have the same format, and it is important to know what area of the curve (or probability) the table presents as corresponding to a given z-score. Table 2 in Appendix B gives the area of the curve lying at or below z. The area to the right of z (or the probability of obtaining a value above z) is simply 1 minus the tabled probability.
- **Using n instead of $n - 1$ degrees of freedom in one-sample t-tests:** Remember that you must subtract 1 from n in order to get the degrees-of-freedom parameter that you need in order to look up a value in the t-table.

- **Confusing confidence level with confidence interval:** The **confidence level** is the significance level of the test or the likelihood of obtaining a given result by chance. The **confidence interval** is a range of values between the lowest and highest values that the estimated parameter could take at a given confidence level.
- **Confusing interval width with margin of error:** A confidence interval is always a point estimate plus or minus a margin of error. The interval width is double that margin of error. If, for example, a population parameter is estimated to be 46 percent plus or minus 4 percent, the interval width is 8 percent.
- **Confusing statistics with parameters:** Parameters are characteristics of the population that you usually do not know; they are designated with Greek symbols (μ and σ). Statistics are characteristics of samples that you are usually able to compute. Although statistics correspond to parameters ($\bar{x}$ is the mean of a sample, as μ is the mean of a population), the two are not interchangeable; hence, you need to be careful and know which variables are parameters and which are statistics. You compute statistics in order to estimate parameters.
- **Confusing the addition rule with the multiplication rule:** When determining probability, the multiplication rule applies if all favorable outcomes must occur in a series of events. The addition rule applies when *at least one* success must occur in a series of events.
- **Forgetting that the addition rule applies to mutually exclusive outcomes:** If outcomes can occur together, the probability of their joint occurrence must be subtracted from the total "addition-rule" probability. (See Chapter 4 for more information.)
- **Forgetting to average the two middle values of an even-numbered set when figuring median and (sometimes) quartiles:** If an ordered series contains an even number of measures, the median is always the mean of the two middle measures.
- **Forgetting where to place the points of a frequency polygon:** The points of a frequency polygon are always at the center of each of the class intervals, not at the ends.

Appendix B
TABLES

This appendix includes the following tables:

Table 1: Binomial Probabilities, $P(x)$ for $n \leq 20$

Table 2: Standard Normal Probabilities

Table 3: t Distribution Critical Values

Table 4: χ^2 Critical Values

Table 1 Binomial Probabilities, $P(x)$ for $n \leq 20$

***n* = 2**

	π										
x↓	***0.05***	***0.10***	***0.15***	***0.20***	***0.25***	***0.30***	***0.35***	***0.40***	***0.45***	***0.50***	
0	0.9025	0.8100	0.7225	0.6400	0.5625	0.4900	0.4225	0.3600	0.3025	0.2500	2
1	0.0950	0.1800	0.2550	0.3200	0.3750	0.4200	0.4550	0.4800	0.4950	0.5000	1
2	0.0025	0.0100	0.0225	0.0400	0.0625	0.0900	0.1225	0.1600	0.2025	0.2500	0
	0.95	0.90	0.85	0.80	0.75	0.70	0.65	0.60	0.55	0.50	*x*↑

***n* = 3**

	π										
x↓	***0.05***	***0.10***	***0.15***	***0.20***	***0.25***	***0.30***	***0.35***	***0.40***	***0.45***	***0.50***	
0	0.8574	0.7290	0.6141	0.5120	0.4219	0.3430	0.2746	0.2160	0.1664	0.1250	3
1	0.1354	0.2430	0.3251	0.3840	0.4219	0.4410	0.4436	0.4320	0.4084	0.3750	2
2	0.0071	0.0270	0.0574	0.0960	0.1406	0.1890	0.2389	0.2880	0.3341	0.3750	1
3	0.0001	0.0010	0.0034	0.0080	0.0156	0.0270	0.0429	0.0640	0.0911	0.1250	0
	0.95	0.90	0.85	0.80	0.75	0.70	0.65	0.60	0.55	0.50	*x*↑

***n* = 4**

					π						
x↓	**0.05**	**0.10**	**0.15**	**0.20**	**0.25**	**0.30**	**0.35**	**0.40**	**0.45**	**0.50**	
0	0.8145	0.6561	0.5220	0.4096	0.3164	0.2401	0.1785	0.1296	0.0915	0.0625	4
1	0.1715	0.2916	0.3685	0.4096	0.4219	0.4116	0.3845	0.3456	0.2995	0.2500	3
2	0.0135	0.0486	0.0975	0.1536	0.2109	0.2646	0.3105	0.3456	0.3675	0.3750	2
3	0.0005	0.0036	0.0115	0.0256	0.0469	0.0756	0.1115	0.1536	0.2005	0.2500	1
4	0.0000	0.0001	0.0005	0.0016	0.0039	0.0081	0.0150	0.0256	0.0410	0.0625	0
	0.95	0.90	0.85	0.80	0.75	0.70	0.65	0.60	0.55	0.50	*x*↑

***n* = 5**

					π						
x↓	**0.05**	**0.10**	**0.15**	**0.20**	**0.25**	**0.30**	**0.35**	**0.40**	**0.45**	**0.50**	
0	0.7738	0.5905	0.4437	0.3277	0.2373	0.1681	0.1160	0.0778	0.0503	0.0313	5
1	0.2036	0.3281	0.3915	0.4096	0.3955	0.3602	0.3124	0.2592	0.2059	0.1563	4
2	0.0214	0.0729	0.1382	0.2048	0.2637	0.3087	0.3364	0.3456	0.3369	0.3125	3
3	0.0011	0.0081	0.0244	0.0512	0.0879	0.1323	0.1811	0.2304	0.2757	0.3125	2
4	0.0000	0.0005	0.0022	0.0064	0.0146	0.0284	0.0488	0.0768	0.1128	0.1563	1
5	0.0000	0.0000	0.0001	0.0003	0.0010	0.0024	0.0053	0.0102	0.0185	0.0313	0
	0.95	0.90	0.85	0.80	0.75	0.70	0.65	0.60	0.55	0.50	*x*↑

n = 6

x↓	**0.05**	**0.10**	**0.15**	**0.20**	π **0.25**	**0.30**	**0.35**	**0.40**	**0.45**	**0.50**	
0	0.7351	0.5314	0.3771	0.2621	0.1780	0.1176	0.0754	0.0467	0.0277	0.0156	6
1	0.2321	0.3543	0.3993	0.3932	0.3560	0.3025	0.2437	0.1866	0.1359	0.0938	5
2	0.0305	0.0984	0.1762	0.2458	0.2966	0.3241	0.3280	0.3110	0.2780	0.2344	4
3	0.0021	0.0146	0.0415	0.0819	0.1318	0.1852	0.2355	0.2765	0.3032	0.3125	3
4	0.0001	0.0012	0.0055	0.0154	0.0330	0.0595	0.0951	0.1382	0.1861	0.2344	2
5	0.0000	0.0001	0.0004	0.0015	0.0044	0.0102	0.0205	0.0369	0.0609	0.0938	1
6	0.0000	0.0000	0.0000	0.0001	0.0002	0.0007	0.0018	0.0041	0.0083	0.0156	0
	0.95	0.90	0.85	0.80	0.75	0.70	0.65	0.60	0.55	0.50	x↑

***n* = 7**

x↓	0.05	0.10	0.15	0.20	π 0.25	0.30	0.35	0.40	0.45	0.50	
0	0.6983	0.4783	0.3206	0.2097	0.1335	0.0824	0.0490	0.0280	0.0152	0.0078	7
1	0.2573	0.3720	0.3960	0.3670	0.3115	0.2471	0.1848	0.1306	0.0872	0.0547	6
2	0.0406	0.1240	0.2097	0.2753	0.3115	0.3177	0.2985	0.2613	0.2140	0.1641	5
3	0.0036	0.0230	0.0617	0.1147	0.1730	0.2269	0.2679	0.2903	0.2918	0.2734	4
4	0.0002	0.0026	0.0109	0.0287	0.0577	0.0972	0.1442	0.1935	0.2388	0.2734	3
5	0.0000	0.0002	0.0012	0.0043	0.0115	0.0250	0.0466	0.0774	0.1172	0.1641	2
6	0.0000	0.0000	0.0001	0.0004	0.0013	0.0036	0.0084	0.0172	0.0320	0.0547	1
7	0.0000	0.0000	0.0000	0.0000	0.0001	0.0002	0.0006	0.0016	0.0037	0.0078	0
	0.95	0.90	0.85	0.80	0.75	0.70	0.65	0.60	0.55	0.50	x↑

***n* = 8**

$x\downarrow$	***0.05***	***0.10***	***0.15***	***0.20***	π ***0.25***	***0.30***	***0.35***	***0.40***	***0.45***	***0.50***	
0	0.6634	0.4305	0.2725	0.1678	0.1001	0.0576	0.0319	0.0168	0.0084	0.0039	8
1	0.2793	0.3826	0.3847	0.3355	0.2670	0.1977	0.1373	0.0896	0.0548	0.0313	7
2	0.0515	0.1488	0.2376	0.2936	0.3115	0.2965	0.2587	0.2090	0.1569	0.1094	6
3	0.0054	0.0331	0.0839	0.1468	0.2076	0.2541	0.2786	0.2787	0.2568	0.2188	5
4	0.0004	0.0046	0.0185	0.0459	0.0865	0.1361	0.1875	0.2322	0.2627	0.2734	4
5	0.0000	0.0004	0.0026	0.0092	0.0231	0.0467	0.0808	0.1239	0.1719	0.2188	3
6	0.0000	0.0000	0.0002	0.0011	0.0038	0.0100	0.0217	0.0413	0.0703	0.1094	2
7	0.0000	0.0000	0.0000	0.0001	0.0004	0.0012	0.0033	0.0079	0.0164	0.0313	1
8	0.0000	0.0000	0.0000	0.0000	0.0000	0.0001	0.0002	0.0007	0.0017	0.0039	0
	0.95	0.90	0.85	0.80	0.75	0.70	0.65	0.60	0.55	0.50	$x\uparrow$

$n = 9$

$x\downarrow$	**0.05**	**0.10**	**0.15**	**0.20**	π **0.25**	**0.30**	**0.35**	**0.40**	**0.45**	**0.50**	
0	0.6302	0.3874	0.2316	0.1342	0.0751	0.0404	0.0207	0.0101	0.0046	0.0020	9
1	0.2985	0.3874	0.3679	0.3020	0.2253	0.1556	0.1004	0.0605	0.0339	0.0176	8
2	0.0629	0.1722	0.2597	0.3020	0.3003	0.2668	0.2162	0.1612	0.1110	0.0703	7
3	0.0077	0.0446	0.1069	0.1762	0.2336	0.2668	0.2716	0.2508	0.2119	0.1641	6
4	0.0006	0.0074	0.0283	0.0661	0.1168	0.1715	0.2194	0.2508	0.2600	0.2461	5
5	0.0000	0.0008	0.0050	0.0165	0.0389	0.0735	0.1181	0.1672	0.2128	0.2461	4
6	0.0000	0.0001	0.0006	0.0028	0.0087	0.0210	0.0424	0.0743	0.1160	0.1641	3
7	0.0000	0.0000	0.0000	0.0003	0.0012	0.0039	0.0098	0.0212	0.0407	0.0703	2
8	0.0000	0.0000	0.0000	0.0000	0.0001	0.0004	0.0013	0.0035	0.0083	0.0176	1
9	0.0000	0.0000	0.0000	0.0000	0.0000	0.0000	0.0001	0.0003	0.0008	0.0020	0
	0.95	0.90	0.85	0.80	0.75	0.70	0.65	0.60	0.55	0.50	$x\uparrow$

***n* = 10**

$x\downarrow$	**0.05**	**0.10**	**0.15**	**0.20**	π **0.25**	**0.30**	**0.35**	**0.40**	**0.45**	**0.50**	
0	0.5987	0.3487	0.1969	0.1074	0.0563	0.0282	0.0135	0.0060	0.0025	0.0010	10
1	0.3151	0.3874	0.3474	0.2684	0.1877	0.1211	0.0725	0.0403	0.0207	0.0098	9
2	0.0746	0.1937	0.2759	0.3020	0.2816	0.2335	0.1757	0.1209	0.0763	0.0439	8
3	0.0105	0.0574	0.1298	0.2013	0.2503	0.2668	0.2522	0.2150	0.1665	0.1172	7
4	0.0010	0.0112	0.0401	0.0881	0.1460	0.2001	0.2377	0.2508	0.2384	0.2051	6
5	0.0001	0.0015	0.0085	0.0264	0.0584	0.1029	0.1536	0.2007	0.2340	0.2461	5
6	0.0000	0.0001	0.0012	0.0055	0.0162	0.0368	0.0689	0.1115	0.1596	0.2051	4
7	0.0000	0.0000	0.0001	0.0008	0.0031	0.0090	0.0212	0.0425	0.0746	0.1172	3
8	0.0000	0.0000	0.0000	0.0001	0.0004	0.0014	0.0043	0.0106	0.0229	0.0439	2
9	0.0000	0.0000	0.0000	0.0000	0.0000	0.0001	0.0005	0.0016	0.0042	0.0098	1
10	0.0000	0.0000	0.0000	0.0000	0.0000	0.0000	0.0000	0.0001	0.0003	0.0010	0
	0.95	0.90	0.85	0.80	0.75	0.70	0.65	0.60	0.55	0.50	$x\uparrow$

***n* = 12**

$x\downarrow$	***0.05***	***0.10***	***0.15***	***0.20***	π ***0.25***	***0.30***	***0.35***	***0.40***	***0.45***	***0.50***	
0	0.5404	0.2824	0.1422	0.0687	0.0317	0.0138	0.0057	0.0022	0.0008	0.0002	12
1	0.3413	0.3766	0.3012	0.2062	0.1267	0.0712	0.0368	0.0174	0.0075	0.0029	11
2	0.0988	0.2301	0.2924	0.2835	0.2323	0.1678	0.1088	0.0639	0.0339	0.0161	10
3	0.0173	0.0852	0.1720	0.2362	0.2581	0.2397	0.1954	0.1419	0.0923	0.0537	9
4	0.0021	0.0213	0.0683	0.1329	0.1936	0.2311	0.2367	0.2128	0.1700	0.1208	8
5	0.0002	0.0038	0.0193	0.0532	0.1032	0.1585	0.2039	0.2270	0.2225	0.1934	7
6	0.0000	0.0005	0.0040	0.0155	0.0401	0.0792	0.1281	0.1766	0.2124	0.2256	6
7	0.0000	0.0000	0.0006	0.0033	0.0115	0.0291	0.0591	0.1009	0.1489	0.1934	5
8	0.0000	0.0000	0.0001	0.0005	0.0024	0.0078	0.0199	0.0420	0.0762	0.1208	4
9	0.0000	0.0000	0.0000	0.0001	0.0004	0.0015	0.0048	0.0125	0.0277	0.0537	3
10	0.0000	0.0000	0.0000	0.0000	0.0000	0.0002	0.0008	0.0025	0.0068	0.0161	2
11	0.0000	0.0000	0.0000	0.0000	0.0000	0.0000	0.0001	0.0003	0.0010	0.0029	1
12	0.0000	0.0000	0.0000	0.0000	0.0000	0.0000	0.0000	0.0000	0.0001	0.0002	0
	0.95	0.90	0.85	0.80	0.75	0.70	0.65	0.60	0.55	0.50	$x\uparrow$

***n* = 14**

x↓					π						
	0.05	***0.10***	***0.15***	***0.20***	***0.25***	***0.30***	***0.35***	***0.40***	***0.45***	***0.50***	
0	0.4877	0.2288	0.1028	0.0440	0.0178	0.0068	0.0024	0.0008	0.0002	0.0001	14
1	0.3593	0.3559	0.2539	0.1539	0.0832	0.0407	0.0181	0.0073	0.0027	0.0009	13
2	0.1229	0.2570	0.2912	0.2501	0.1802	0.1134	0.0634	0.0317	0.0141	0.0056	12
3	0.0259	0.1142	0.2056	0.2501	0.2402	0.1943	0.1366	0.0845	0.0462	0.0222	11
4	0.0037	0.0349	0.0998	0.1720	0.2202	0.2290	0.2022	0.1549	0.1040	0.0611	10
5	0.0004	0.0078	0.0352	0.0860	0.1468	0.1963	0.2178	0.2066	0.1701	0.1222	9
6	0.0000	0.0013	0.0093	0.0322	0.0734	0.1262	0.1759	0.2066	0.2088	0.1833	8
7	0.0000	0.0002	0.0019	0.0092	0.0280	0.0618	0.1082	0.1574	0.1952	0.2095	7
8	0.0000	0.0000	0.0003	0.0020	0.0082	0.0232	0.0510	0.0918	0.1398	0.1833	6
9	0.0000	0.0000	0.0000	0.0003	0.0018	0.0066	0.0183	0.0408	0.0762	0.1222	5
10	0.0000	0.0000	0.0000	0.0000	0.0003	0.0014	0.0049	0.0136	0.0312	0.0611	4
11	0.0000	0.0000	0.0000	0.0000	0.0000	0.0002	0.0010	0.0033	0.0093	0.0222	3
12	0.0000	0.0000	0.0000	0.0000	0.0000	0.0000	0.0001	0.0005	0.0019	0.0056	2
13	0.0000	0.0000	0.0000	0.0000	0.0000	0.0000	0.0000	0.0001	0.0002	0.0009	1
14	0.0000	0.0000	0.0000	0.0000	0.0000	0.0000	0.0000	0.0000	0.0000	0.0001	0
	0.95	0.90	0.85	0.80	0.75	0.70	0.65	0.60	0.55	0.50	*x*↑

***n* = 16**

					π						
x↓	**0.05**	**0.10**	**0.15**	**0.20**	**0.25**	**0.30**	**0.35**	**0.40**	**0.45**	**0.50**	
0	0.4401	0.1853	0.0743	0.0281	0.0100	0.0033	0.0010	0.0003	0.0001	0.0000	16
1	0.3706	0.3294	0.2097	0.1126	0.0535	0.0228	0.0087	0.0030	0.0009	0.0002	15
2	0.1463	0.2745	0.2775	0.2111	0.1336	0.0732	0.0353	0.0150	0.0056	0.0018	14
3	0.0359	0.1423	0.2285	0.2463	0.2079	0.1465	0.0888	0.0468	0.0215	0.0085	13
4	0.0061	0.0514	0.1311	0.2001	0.2252	0.2040	0.1553	0.1014	0.0572	0.0278	12
5	0.0008	0.0137	0.0555	0.1201	0.1802	0.2099	0.2008	0.1623	0.1123	0.0667	11
6	0.0001	0.0028	0.0180	0.0550	0.1101	0.1649	0.1982	0.1983	0.1684	0.1222	10
7	0.0000	0.0004	0.0045	0.0197	0.0524	0.1010	0.1524	0.1889	0.1969	0.1746	9
8	0.0000	0.0001	0.0009	0.0055	0.0197	0.0487	0.0923	0.1417	0.1812	0.1964	8
9	0.0000	0.0000	0.0001	0.0012	0.0058	0.0185	0.0442	0.0840	0.1318	0.1746	7
10	0.0000	0.0000	0.0000	0.0002	0.0014	0.0056	0.0167	0.0392	0.0755	0.1222	6
11	0.0000	0.0000	0.0000	0.0000	0.0002	0.0013	0.0049	0.0142	0.0337	0.0667	5
12	0.0000	0.0000	0.0000	0.0000	0.0000	0.0002	0.0011	0.0040	0.0115	0.0278	4
13	0.0000	0.0000	0.0000	0.0000	0.0000	0.0000	0.0002	0.0008	0.0029	0.0085	3
14	0.0000	0.0000	0.0000	0.0000	0.0000	0.0000	0.0000	0.0001	0.0005	0.0018	2
15	0.0000	0.0000	0.0000	0.0000	0.0000	0.0000	0.0000	0.0000	0.0001	0.0002	1
	0.95	0.90	0.85	0.80	0.75	0.70	0.65	0.60	0.55	0.50	x↑

***n* = 18**

$x\downarrow$	**0.05**	**0.10**	**0.15**	**0.20**	π **0.25**	**0.30**	**0.35**	**0.40**	**0.45**	**0.50**	
0	0.3972	0.1501	0.0536	0.0180	0.0056	0.0016	0.0004	0.0001	0.0000	0.0000	18
1	0.3763	0.3002	0.1704	0.0811	0.0338	0.0126	0.0042	0.0012	0.0003	0.0001	17
2	0.1683	0.2835	0.2556	0.1723	0.0958	0.0458	0.0190	0.0069	0.0022	0.0006	16
3	0.0473	0.1680	0.2406	0.2297	0.1704	0.1046	0.0547	0.0246	0.0095	0.0031	15
4	0.0093	0.0700	0.1592	0.2153	0.2130	0.1681	0.1104	0.0614	0.0291	0.0117	14
5	0.0014	0.0218	0.0787	0.1507	0.1988	0.2017	0.1664	0.1146	0.0666	0.0327	13
6	0.0002	0.0052	0.0301	0.0816	0.1436	0.1873	0.1941	0.1655	0.1181	0.0708	12
7	0.0000	0.0010	0.0091	0.0350	0.0820	0.1376	0.1792	0.1892	0.1657	0.1214	11
8	0.0000	0.0002	0.0022	0.0120	0.0376	0.0811	0.1327	0.1734	0.1864	0.1669	10
9	0.0000	0.0000	0.0004	0.0033	0.0139	0.0386	0.0794	0.1284	0.1694	0.1855	9
10	0.0000	0.0000	0.0001	0.0008	0.0042	0.0149	0.0385	0.0771	0.1248	0.1669	8
11	0.0000	0.0000	0.0000	0.0001	0.0010	0.0046	0.0151	0.0374	0.0742	0.1214	7
12	0.0000	0.0000	0.0000	0.0000	0.0002	0.0012	0.0047	0.0145	0.0354	0.0708	6
13	0.0000	0.0000	0.0000	0.0000	0.0000	0.0002	0.0012	0.0045	0.0134	0.0327	5
14	0.0000	0.0000	0.0000	0.0000	0.0000	0.0000	0.0002	0.0011	0.0039	0.0117	4
15	0.0000	0.0000	0.0000	0.0000	0.0000	0.0000	0.0000	0.0002	0.0009	0.0031	3
16	0.0000	0.0000	0.0000	0.0000	0.0000	0.0000	0.0000	0.0000	0.0001	0.0006	2
17	0.0000	0.0000	0.0000	0.0000	0.0000	0.0000	0.0000	0.0000	0.0000	0.0001	1
	0.95	0.90	0.85	0.80	0.75	0.70	0.65	0.60	0.55	0.50	$x\uparrow$

***n* = 20**

					π						
x↓	***0.05***	***0.10***	***0.15***	***0.20***	***0.25***	***0.30***	***0.35***	***0.40***	***0.45***	***0.50***	
0	0.3585	0.1216	0.0388	0.0115	0.0032	0.0008	0.0002	0.0000	0.0000	0.0000	20
1	0.3774	0.2702	0.1368	0.0576	0.0211	0.0068	0.0020	0.0005	0.0001	0.0000	19
2	0.1887	0.2852	0.2293	0.1369	0.0669	0.0278	0.0100	0.0031	0.0008	0.0002	18
3	0.0596	0.1901	0.2428	0.2054	0.1339	0.0716	0.0323	0.0123	0.0040	0.0011	17
4	0.0133	0.0898	0.1821	0.2182	0.1897	0.1304	0.0738	0.0350	0.0139	0.0046	16
5	0.0022	0.0319	0.1028	0.1746	0.2023	0.1789	0.1272	0.0746	0.0365	0.0148	15
6	0.0003	0.0089	0.0454	0.1091	0.1686	0.1916	0.1712	0.1244	0.0746	0.0370	14
7	0.0000	0.0020	0.0160	0.0545	0.1124	0.1643	0.1844	0.1659	0.1221	0.0739	13
8	0.0000	0.0004	0.0046	0.0222	0.0609	0.1144	0.1614	0.1797	0.1623	0.1201	12
9	0.0000	0.0001	0.0011	0.0074	0.0271	0.0654	0.1158	0.1597	0.1771	0.1602	11
10	0.0000	0.0000	0.0002	0.0020	0.0099	0.0308	0.0686	0.1171	0.1593	0.1762	10
11	0.0000	0.0000	0.0000	0.0005	0.0030	0.0120	0.0336	0.0710	0.1185	0.1602	9
12	0.0000	0.0000	0.0000	0.0001	0.0008	0.0039	0.0136	0.0355	0.0727	0.1201	8
13	0.0000	0.0000	0.0000	0.0000	0.0002	0.0010	0.0045	0.0146	0.0366	0.0739	7
14	0.0000	0.0000	0.0000	0.0000	0.0000	0.0002	0.0012	0.0049	0.0150	0.0370	6
15	0.0000	0.0000	0.0000	0.0000	0.0000	0.0000	0.0003	0.0013	0.0049	0.0148	5
16	0.0000	0.0000	0.0000	0.0000	0.0000	0.0000	0.0000	0.0003	0.0013	0.0046	4
17	0.0000	0.0000	0.0000	0.0000	0.0000	0.0000	0.0000	0.0000	0.0002	0.0011	3
18	0.0000	0.0000	0.0000	0.0000	0.0000	0.0000	0.0000	0.0000	0.0000	0.0002	2
	0.95	0.90	0.85	0.80	0.75	0.70	0.65	0.60	0.55	0.50	x↑

Table 2 Standard Normal Probabilities

(Table entry is probability at or below *z*.)

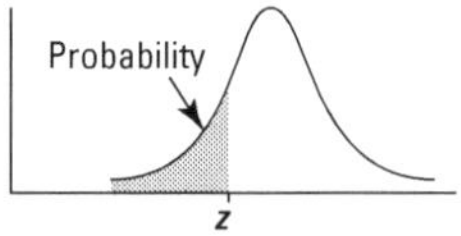

z	0.00	0.01	0.02	0.03	0.04
−3.4	0.0003	0.0003	0.0003	0.0003	0.0003
−3.3	0.0005	0.0005	0.0005	0.0004	0.0004
−3.2	0.0007	0.0007	0.0006	0.0006	0.0006
−3.1	0.0010	0.0009	0.0009	0.0009	0.0008
−3.0	0.0013	0.0013	0.0013	0.0012	0.0012
−2.9	0.0019	0.0018	0.0018	0.0017	0.0016
−2.8	0.0026	0.0025	0.0024	0.0023	0.0023
−2.7	0.0035	0.0034	0.0033	0.0032	0.0031
−2.6	0.0047	0.0045	0.0044	0.0043	0.0041
−2.5	0.0062	0.0060	0.0059	0.0057	0.0055
−2.4	0.0082	0.0080	0.0078	0.0075	0.0073
−2.3	0.0107	0.0104	0.0102	0.0099	0.0096
−2.2	0.0139	0.0136	0.0132	0.0129	0.0125
−2.1	0.0179	0.0174	0.0170	0.0166	0.0162
−2.0	0.0228	0.0222	0.0217	0.0212	0.0207
−1.9	0.0287	0.0281	0.0274	0.0268	0.0262
−1.8	0.0359	0.0351	0.0344	0.0336	0.0329
−1.7	0.0446	0.0436	0.0427	0.0418	0.0409
−1.6	0.0548	0.0537	0.0526	0.0516	0.0505
−1.5	0.0668	0.0655	0.0643	0.0630	0.0618
−1.4	0.0808	0.0793	0.0778	0.0764	0.0749
−1.3	0.0968	0.0951	0.0934	0.0918	0.0901
−1.2	0.1151	0.1131	0.1112	0.1093	0.1075
−1.1	0.1357	0.1335	0.1314	0.1292	0.1271
−1.0	0.1587	0.1562	0.1539	0.1515	0.1492
−0.9	0.1841	0.1814	0.1788	0.1762	0.1736
−0.8	0.2119	0.2090	0.2061	0.2033	0.2005
−0.7	0.2420	0.2389	0.2358	0.2327	0.2296
−0.6	0.2743	0.2709	0.2676	0.2643	0.2611
−0.5	0.3085	0.3050	0.3015	0.2981	0.2946
−0.4	0.3446	0.3409	0.3372	0.3336	0.3300
−0.3	0.3821	0.3783	0.3745	0.3707	0.3669
−0.2	0.4207	0.4168	0.4129	0.4090	0.4052
−0.1	0.4602	0.4562	0.4522	0.4483	0.4443
−0.0	0.5000	0.4960	0.4920	0.4880	0.4840

0.05	0.06	0.07	0.08	0.09	z
0.0003	0.0003	0.0003	0.0003	0.0002	–3.4
0.0004	0.0004	0.0004	0.0004	0.0003	–3.3
0.0006	0.0006	0.0005	0.0005	0.0005	–3.2
0.0008	0.0008	0.0008	0.0007	0.0007	–3.1
0.0011	0.0011	0.0011	0.0010	0.0010	–3.0
0.0016	0.0015	0.0015	0.0014	0.0014	–2.9
0.0022	0.0021	0.0021	0.0020	0.0019	–2.8
0.0030	0.0029	0.0028	0.0027	0.0026	–2.7
0.0040	0.0039	0.0038	0.0037	0.0036	–2.6
0.0054	0.0052	0.0051	0.0049	0.0048	–2.5
0.0071	0.0069	0.0068	0.0066	0.0064	–2.4
0.0094	0.0091	0.0089	0.0087	0.0084	–2.3
0.0122	0.0119	0.0116	0.0113	0.0110	–2.2
0.0158	0.0154	0.0150	0.0146	0.0143	–2.1
0.0202	0.0197	0.0192	0.0188	0.0183	–2.0
0.0256	0.0250	0.0244	0.0239	0.0233	–1.9
0.0322	0.0314	0.0307	0.0301	0.0294	–1.8
0.0401	0.0392	0.0384	0.0375	0.0367	–1.7
0.0495	0.0485	0.0475	0.0465	0.0455	–1.6
0.0606	0.0594	0.0582	0.0571	0.0559	–1.5
0.0735	0.0721	0.0708	0.0694	0.0681	–1.4
0.0885	0.0869	0.0853	0.0838	0.0823	–1.3
0.1056	0.1038	0.1020	0.1003	0.0985	–1.2
0.1251	0.1230	0.1210	0.1190	0.1170	–1.1
0.1469	0.1446	0.1423	0.1401	0.1379	–1.0
0.1711	0.1685	0.1660	0.1635	0.1611	–0.9
0.1977	0.1949	0.1922	0.1894	0.1867	–0.8
0.2266	0.2236	0.2206	0.2177	0.2148	–0.7
0.2578	0.2546	0.2514	0.2483	0.2451	–0.6
0.2912	0.2877	0.2843	0.2810	0.2776	–0.5
0.3264	0.3228	0.3192	0.3156	0.3121	–0.4
0.3632	0.3594	0.3557	0.3520	0.3483	–0.3
0.4013	0.3974	0.3936	0.3897	0.3859	–0.2
0.4404	0.4364	0.4325	0.4286	0.4247	–0.1
0.4801	0.4761	0.4721	0.4681	0.4641	–0.0

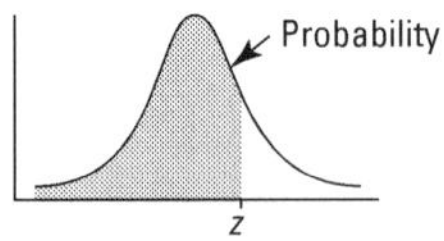

z	0.00	0.01	0.02	0.03	0.04
0.0	0.5000	0.5040	0.5080	0.5120	0.5160
0.1	0.5398	0.5438	0.5478	0.5517	0.5557
0.2	0.5793	0.5832	0.5871	0.5910	0.5948
0.3	0.6179	0.6217	0.6255	0.6293	0.6331
0.4	0.6554	0.6591	0.6628	0.6664	0.6700
0.5	0.6915	0.6950	0.6985	0.7019	0.7054
0.6	0.7257	0.7291	0.7324	0.7357	0.7389
0.7	0.7580	0.7611	0.7642	0.7673	0.7704
0.8	0.7881	0.7910	0.7939	0.7967	0.7995
0.9	0.8159	0.8186	0.8212	0.8238	0.8264
1.0	0.8413	0.8438	0.8461	0.8485	0.8508
1.1	0.8643	0.8665	0.8686	0.8708	0.8729
1.2	0.8849	0.8869	0.8888	0.8907	0.8925
1.3	0.9032	0.9049	0.9066	0.9082	0.9099
1.4	0.9192	0.9207	0.9222	0.9236	0.9251
1.5	0.9332	0.9345	0.9357	0.9370	0.9382
1.6	0.9452	0.9463	0.9474	0.9484	0.9495
1.7	0.9554	0.9564	0.9573	0.9582	0.9591
1.8	0.9641	0.9649	0.9656	0.9664	0.9671
1.9	0.9713	0.9719	0.9726	0.9732	0.9738
2.0	0.9772	0.9778	0.9783	0.9788	0.9793
2.1	0.9821	0.9826	0.9830	0.9834	0.9838
2.2	0.9861	0.9864	0.9868	0.9871	0.9875
2.3	0.9893	0.9896	0.9898	0.9901	0.9904
2.4	0.9918	0.9920	0.9922	0.9925	0.9927
2.5	0.9938	0.9940	0.9941	0.9943	0.9945
2.6	0.9953	0.9955	0.9956	0.9957	0.9959
2.7	0.9965	0.9966	0.9967	0.9968	0.9969
2.8	0.9974	0.9975	0.9976	0.9977	0.9977
2.9	0.9981	0.9982	0.9982	0.9983	0.9984
3.0	0.9987	0.9987	0.9987	0.9988	0.9988
3.1	0.9990	0.9991	0.9991	0.9991	0.9992
3.2	0.9993	0.9993	0.9994	0.9994	0.9994
3.3	0.9995	0.9995	0.9995	0.9996	0.9996
3.4	0.9997	0.9997	0.9997	0.9997	0.9997

0.05	***0.06***	***0.07***	***0.08***	***0.09***	***z***
0.5199	0.5239	0.5279	0.5319	0.5359	0.0
0.5596	0.5636	0.5675	0.5714	0.5753	0.1
0.5987	0.6026	0.6064	0.6103	0.6141	0.2
0.6368	0.6406	0.6443	0.6480	0.6517	0.3
0.6736	0.6772	0.6808	0.6844	0.6879	0.4
0.7088	0.7123	0.7157	0.7190	0.7224	0.5
0.7422	0.7454	0.7486	0.7517	0.7549	0.6
0.7734	0.7764	0.7794	0.7823	0.7852	0.7
0.8023	0.8051	0.8078	0.8106	0.8133	0.8
0.8289	0.8315	0.8340	0.8365	0.8389	0.9
0.8531	0.8554	0.8577	0.8599	0.8621	1.0
0.8749	0.8770	0.8790	0.8810	0.8830	1.1
0.8944	0.8962	0.8980	0.8997	0.9015	1.2
0.9115	0.9131	0.9147	0.9162	0.9177	1.3
0.9265	0.9279	0.9292	0.9306	0.9319	1.4
0.9394	0.9406	0.9418	0.9429	0.9441	1.5
0.9505	0.9515	0.9525	0.9535	0.9545	1.6
0.9599	0.9608	0.9616	0.9625	0.9633	1.7
0.9678	0.9686	0.9693	0.9699	0.9706	1.8
0.9744	0.9750	0.9756	0.9761	0.9767	1.9
0.9798	0.9803	0.9808	0.9812	0.9817	2.0
0.9842	0.9846	0.9850	0.9854	0.9857	2.1
0.9878	0.9881	0.9884	0.9887	0.9890	2.2
0.9906	0.9909	0.9911	0.9913	0.9916	2.3
0.9929	0.9931	0.9932	0.9934	0.9936	2.4
0.9946	0.9948	0.9949	0.9951	0.9952	2.5
0.9960	0.9961	0.9962	0.9963	0.9964	2.6
0.9970	0.9971	0.9972	0.9973	0.9974	2.7
0.9978	0.9979	0.9979	0.9980	0.9981	2.8
0.9984	0.9985	0.9985	0.9986	0.9986	2.9
0.9989	0.9989	0.9989	0.9990	0.9990	3.0
0.9992	0.9992	0.9992	0.9993	0.9993	3.1
0.9994	0.9994	0.9995	0.9995	0.9995	3.2
0.9996	0.9996	0.9996	0.9996	0.9997	3.3
0.9997	0.9997	0.9997	0.9997	0.9998	3.4

Table 3 *t* Distribution Critical Values

(Table entry is the point t^* with given probability p lying above it.)

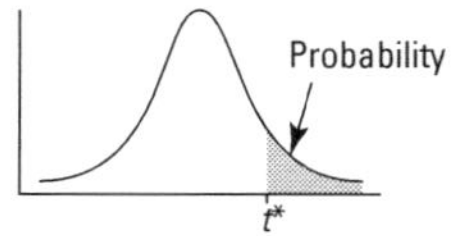

	p											
df	***0.25***	***0.20***	***0.15***	***0.10***	***0.05***	***0.025***	***0.02***	***0.01***	***0.005***	***0.0025***	***0.001***	***0.0005***
1	1.000	1.376	1.963	3.078	6.314	12.71	15.89	31.82	63.66	127.3	318.3	636.6
2	0.816	1.061	1.386	1.886	2.920	4.303	4.849	6.965	9.925	14.09	22.33	31.60
3	0.765	0.978	1.250	1.638	2.353	3.182	3.482	4.541	5.841	7.453	10.21	12.92
4	0.741	0.941	1.190	1.533	2.132	2.776	2.999	3.747	4.604	5.598	7.173	8.610
5	0.727	0.920	1.156	1.476	2.015	2.571	2.757	3.365	4.032	4.773	5.893	6.869
6	0.718	0.906	1.134	1.440	1.943	2.447	2.612	3.143	3.707	4.317	5.208	5.959
7	0.711	0.896	1.119	1.415	1.895	2.365	2.517	2.998	3.499	4.029	4.785	5.408
8	0.706	0.889	1.108	1.397	1.860	2.306	2.449	2.896	3.355	3.833	4.501	5.041
9	0.703	0.883	1.100	1.383	1.833	2.262	2.398	2.821	3.250	3.690	4.297	4.781
10	0.700	0.879	1.093	1.372	1.812	2.228	2.359	2.764	3.169	3.581	4.144	4.587
11	0.697	0.876	1.088	1.363	1.796	2.201	2.328	2.718	3.106	3.497	4.025	4.437
12	0.695	0.873	1.083	1.356	1.782	2.179	2.303	2.681	3.055	3.428	3.930	4.318
13	0.694	0.870	1.079	1.350	1.771	2.160	2.282	2.650	3.012	3.372	3.852	4.221
14	0.692	0.868	1.076	1.345	1.761	2.145	2.264	2.624	2.977	3.326	3.787	4.140

15	0.691	0.866	1.074	1.341	1.753	2.131	2.249	2.602	2.947	3.286	3.733	4.073
16	0.690	0.865	1.071	1.337	1.746	2.120	2.235	2.583	2.921	3.252	3.686	4.015
17	0.689	0.863	1.069	1.333	1.740	2.110	2.224	2.567	2.898	3.222	3.646	3.965
18	0.688	0.862	1.067	1.330	1.734	2.101	2.214	2.552	2.878	3.197	3.610	3.922
19	0.688	0.861	1.066	1.328	1.729	2.093	2.205	2.539	2.861	3.174	3.579	3.883
20	0.687	0.860	1.064	1.325	1.725	2.086	2.197	2.528	2.845	3.153	3.552	3.850
21	0.686	0.859	1.063	1.323	1.721	2.080	2.189	2.518	2.831	3.135	3.527	3.819
22	0.686	0.858	1.061	1.321	1.717	2.074	2.183	2.508	2.819	3.119	3.505	3.792
23	0.685	0.858	1.060	1.319	1.714	2.069	2.177	2.500	2.807	3.104	3.485	3.768
24	0.685	0.857	1.059	1.318	1.711	2.064	2.172	2.492	2.797	3.091	3.467	3.745
25	0.684	0.856	1.058	1.316	1.708	2.060	2.167	2.485	2.787	3.078	3.450	3.725
26	0.684	0.856	1.058	1.315	1.706	2.056	2.162	2.479	2.779	3.067	3.435	3.707
27	0.684	0.855	1.057	1.314	1.703	2.052	2.158	2.473	2.771	3.057	3.421	3.690
28	0.683	0.855	1.056	1.313	1.701	2.048	2.154	2.467	2.763	3.047	3.408	3.674
29	0.683	0.854	1.055	1.311	1.699	2.045	2.150	2.462	2.756	3.038	3.396	3.659
30	0.683	0.854	1.055	1.310	1.697	2.042	2.147	2.457	2.750	3.030	3.385	3.646
40	0.681	0.851	1.050	1.303	1.684	2.021	2.123	2.423	2.704	2.971	3.307	3.551
50	0.679	0.849	1.047	1.299	1.676	2.009	2.109	2.403	2.678	2.937	3.261	3.496
60	0.679	0.848	1.045	1.296	1.671	2.000	2.099	2.390	2.660	2.915	3.232	3.460
80	0.678	0.846	1.043	1.292	1.664	1.990	2.088	2.374	2.639	2.887	3.195	3.416
100	0.677	0.845	1.042	1.290	1.660	1.984	2.081	2.364	2.626	2.871	3.174	3.390
1000	0.675	0.842	1.037	1.282	1.646	1.962	2.056	2.330	2.581	2.813	3.098	3.300
∞	0.674	0.841	1.036	1.282	1.645	1.960	2.054	2.326	2.576	2.807	3.091	3.291

Table 4 χ^2 Critical Values

(Table entry is the point χ^2 with given probability p lying above it.)

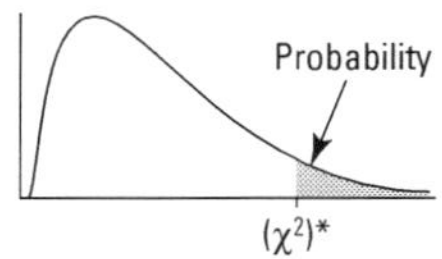

df	*p*											
	0.25	**0.20**	**0.15**	**0.10**	**0.05**	**0.025**	**0.02**	**0.01**	**0.005**	**0.0025**	**0.001**	**0.0005**
1	1.32	1.64	2.07	2.71	3.84	5.02	5.41	6.63	7.88	9.14	10.83	12.12
2	2.77	3.22	3.79	4.61	5.99	7.38	7.82	9.21	10.60	11.98	13.82	15.20
3	4.11	4.64	5.32	6.25	7.81	9.35	9.84	11.34	12.84	14.32	16.27	17.73
4	5.39	5.99	6.74	7.78	9.49	11.14	11.67	13.28	14.86	16.42	18.47	20.00
5	6.63	7.29	8.12	9.24	11.07	12.83	13.39	15.09	16.75	18.39	20.52	22.11
6	7.84	8.56	9.45	10.64	12.59	14.45	15.03	16.81	18.55	20.25	22.46	24.10
7	9.04	9.80	10.75	12.02	14.07	16.01	16.62	18.48	20.28	22.04	24.32	26.02
8	10.22	11.03	12.03	13.36	15.51	17.53	18.17	20.09	21.95	23.77	26.12	27.87
9	11.39	12.24	13.29	14.68	16.92	19.02	19.68	21.67	23.59	25.46	27.88	29.67
10	12.55	13.44	14.53	15.99	18.31	20.48	21.16	23.21	25.19	27.11	29.59	31.42
11	13.70	14.63	15.77	17.28	19.68	21.92	22.62	24.72	26.76	28.73	31.26	33.14
12	14.85	15.81	16.99	18.55	21.03	23.34	24.05	26.22	28.30	30.32	32.91	34.82
13	15.98	16.98	18.20	19.81	22.36	24.74	25.47	27.69	29.82	31.88	34.53	36.48
14	17.12	18.15	19.41	21.06	23.68	26.12	26.87	29.14	31.32	33.43	36.12	38.11

15	18.25	19.31	20.60	22.31	25.00	27.49	28.26	30.58	32.80	34.95	37.70	39.72
16	19.37	20.47	21.79	23.54	26.30	28.85	29.63	32.00	34.27	36.46	39.25	41.31
17	20.49	21.61	22.98	24.77	27.59	30.19	31.00	33.41	35.72	37.95	40.79	42.88
18	21.60	22.76	24.16	25.99	28.87	31.53	32.35	34.81	37.16	39.42	42.31	44.43
19	22.72	23.90	25.33	27.20	30.14	32.85	33.69	36.19	38.58	40.88	43.82	45.97
20	23.83	25.04	26.50	28.41	31.41	34.17	35.02	37.57	40.00	42.34	45.31	47.50
21	24.93	26.17	27.66	29.62	32.67	35.48	36.34	38.93	41.40	43.78	46.80	49.01
22	26.04	27.30	28.82	30.81	33.92	36.78	37.66	40.29	42.80	45.20	48.27	50.51
23	27.14	28.43	29.98	32.01	35.17	38.08	38.97	41.64	44.18	46.62	49.73	52.00
24	28.24	29.55	31.13	33.20	36.42	39.36	40.27	42.98	45.56	48.03	51.18	53.48
25	29.34	30.68	32.28	34.38	37.65	40.65	41.57	44.31	46.93	49.44	52.62	54.95
26	30.43	31.79	33.43	35.56	38.89	41.92	42.86	45.64	48.29	50.83	54.05	56.41
27	31.53	32.91	34.57	36.74	40.11	43.19	44.14	46.96	49.64	52.22	55.48	57.86
28	32.62	34.03	35.71	37.92	41.34	44.46	45.42	48.28	50.99	53.59	56.89	59.30
29	33.71	35.14	36.85	39.09	42.56	45.72	46.69	49.59	52.34	54.97	58.30	60.73
30	34.80	36.25	37.99	40.26	43.77	46.98	47.96	50.89	53.67	56.33	59.70	62.16
40	45.62	47.27	49.24	51.81	55.76	59.34	60.44	63.69	66.77	69.70	73.40	76.09
50	56.33	58.16	60.35	63.17	67.50	71.42	72.61	76.15	79.49	82.66	86.66	89.56
60	66.98	68.97	71.34	74.40	79.08	83.30	84.58	88.38	91.95	95.34	99.61	102.7
80	88.13	90.41	93.11	96.58	101.9	106.6	108.1	112.3	116.3	120.1	124.8	128.3
100	109.1	111.7	114.7	118.5	124.3	129.6	131.1	135.8	140.2	144.3	149.4	153.2

Index

T

U

V

W

X

Z

Made in the USA
Monee, IL
25 October 2023

45216671R00098